MAR 0 3 2017

Withdrawn/ABCL

D0765292

Withdrawn/ABCL

CAT
WARS

WARS

The Devastating Consequences
of a Cuddly Killer

PETER P. MARRA

AND CHRIS SANTELLA

PRINCETON UNIVERSITY PRESS
PRINCETON AND OXFORD

3 9075 05033948 7

Copyright © 2016 by Peter P. Marra and Chris Santella

Requests for permission to reproduce material from this
work should be sent to Permissions, Princeton University Press

Published by Princeton University Press,
41 William Street, Princeton, New Jersey 08540
In the United Kingdom: Princeton University Press,
6 Oxford Street, Woodstock, Oxfordshire OX20 1TR
press.princeton.edu

Jacket design and lettering by Amanda Weiss
Jacket image courtesy of iStock

All Rights Reserved

ISBN 978-0-691-16741-1

British Library Cataloging-in-Publication Data is available

This book has been composed in Sabon LT Std

Printed on acid-free paper. ∞

Printed in the United States of America

10 9 8 7 6 5 4 3 2 1

To my wife, Anne, for her encouragement, and thoughtful support, and to my kids, Aline and Gabe, may they have the opportunities to experience and understand the brilliance of the natural world. To my brother Michael, who loved all animals but left this world too soon.

—P. M.

To my wife, Deidre, for her constant support . . . and to my daughters, Cassidy and Annabel, that they might grow up to find a world still rich in biodiversity.

—C. S.

CONTENTS

CAT
WARS

CHAPTER ONE

The Obituary of the Stephens Island Wren

No matter how much the cats fight, there always
seem to be plenty of kittens.

—Abraham Lincoln

Rising high from the Marlborough Sound into the Maori sky off
the South Island of New Zealand is Stephens Island. The island lies
two miles from the mainland and fifty-five miles from Wellington,
is no bigger than a quarter of a square mile, is taller than it is wide,
and has peaks reaching almost 1,000 feet. Like all of the islands of
the region, Stephens Island has short, craggy, and almost impene-
trable vegetation, likely because the land is guilty of trying to stop
the strong and persistent southeasterly winds sweeping in from
the Antarctic Continent. According to historical accounts, landing
on the island was so treacherous that few people had ever stepped
foot on its shores, which left it largely pristine. In fact, the island
had likely stood in place for millions of years without human impact; if the Maori people had ever visited, they left no trace. Anglo
explorations of the island began in the 1870s, led by New Zealand
maritime officials who had determined that a lighthouse installation was needed to ensure safe passage through nearby channels. Several hundred people had lost their lives in three major

shipwrecks in the mid-1800s in New Zealand, so lighthouse construction had become a priority. By the early 1890s a lighthouse and several modest homes had been erected on Stephens Island for three lighthouse keepers and their families to share. With little human companionship, lighthouse keepers would often bring cats with them to their island outposts. As one story goes, a cat, possibly named Tibbles, made it to Stephens Island and was allowed to roam free.

David Lyall liked his solitude. He was literate, fit, and orderly, but, most important, he could keep a paraffin lantern burning cleanly. Lyall was cautiously excited about his new position as an assistant lighthouse keeper for New Zealand Maritime. It was January 1894, and he would be one of seventeen people at this new outpost. Being a lighthouse keeper in the late 1800s was not an easy job, although the primary duty—keeping the light burning bright and clean—was straightforward, requiring a constant trimming of the wick to maximize the flame and reduce the smoke. Many lives were in the keeper's hands: A rock near an island on a coal-black night could tear through the wooden belly of a ship in a matter of seconds, meaning an almost certain death for sailors. Many sailors at the time could not swim and, ironically, hated water—especially the frigid subantarctic waters enveloping the southern islands of New Zealand.

The challenge of being a lighthouse keeper was one of endurance—enduring rough weather, the claustrophobically small community, the lack of fresh food, and most of all the isolation. Lighthouse keepers received new provisions from the mainland only twice a month. To bolster their larder, they might keep cows for fresh milk, sheep for wool, chickens for eggs. They might garden a bit if the soil and weather permitted. Lyall was not daunted by these challenges. He had a wife and at least one son, so he needed to make a fair wage. He was also eager to pursue his passions, even if doing so meant life on an isolated island. Lyall loved animals and had an insatiable curiosity for natural history and especially for watching birds. An amateur ornithologist, he was especially eager to study the avifauna of the island and envisioned perhaps even preparing some bird specimens for museums.

Lyall likely was preoccupied at an early age with a need for order, and this would have contributed to a need to name and classify everything he saw in nature. In his passion for animals and their nomenclature, Lyall probably found comfort and an explanation for how nature worked. It was like solving a puzzle, revealing something previously unknown, and providing order to a natural landscape that appeared to be in disorder. Lyall was a self-taught naturalist, drawing most of his understanding from the few books that were available on the natural history of New Zealand (field guides had yet to be invented). He may have found inner peace in nature, through watching birds and their behaviors, and identifying and ascribing names to existing species. Lyall also probably found an inherent value in nature that he could not quantify, justify, or even articulate. He was blindly focused on getting to his new post on the largely unexplored and uninhabited island, a place where he could finally pursue his passions. He envisioned spending long nights deep in thought, identifying specimens of plants, insects, and birds, burning through large amounts of paraffin oil— all while supporting his family.

Lyall would find the perfect study species for his avian interests in the Stephens Island Wren, then undescribed but eventually to be named *Xenicus (Traversia) lyalli*. Except for feathers and eggs, the Stephens Island Wren bore more resemblance to a mouse than a bird. It lived a hobbit-like existence, foraging in logs and even in underground burrows and boulder piles. Some accounts even suggest the wren was semi-nocturnal. Equipped with large feet and a short tail, it ran low to the ground among the shoreline rocks or jumped from branch to branch through thick tangles of knotty shrubs. It flapped its vestigial wings to help on the occasional long jump—perhaps its closest approximation to flight. Nearly everything about this species made it wren-like, though it was not actually a member of the wren family (we will continue to refer to it as a wren), but instead was a member of the endemic New Zealand family Acanthisittidae. It was one of only three flightless species of songbirds in the world. It did not need to fly. There was no need to leave the island or the ground for long—food was available throughout the year, and the species could breed on the island.

More important, there were no predators. Flying requires trade-offs with other costly adaptations, and because there was no need to escape or migrate, this small wren, weighing little more than a large coin, lost its ability to fly.

The Stephens Island Wren was millions of years in the making. Enormous evolutionary changes in natural history and biology had occurred over time, generation after generation, to make it unique. Each year wrens nested, laid eggs, raised young—sometimes more, sometimes less, depending on the quality of the mate, the amount of food available, the climate, or some complex mix of all these things. The species' size, color, and shape changed at varying speeds over time, sometimes at a glacial pace, sometimes more rapidly. But this species, like all species of plants and animals, changed at a pace to fit the landscape—the biological, climatic, and geological landscape—all through the process of natural selection. A story told over and over again, all over the planet, for thousands, hundreds of thousands, even hundreds of millions of years. In contrast with the slow pace with which the process of speciation can proceed, the reverse process of species extinction can occur with astonishing speed.

🐾

New Zealand itself is a nation of islands, an archipelago made up of two large landmasses, the South Island and the North Island, surrounded by an array of smaller islands and all isolated from the rest of the world for over 80 million years. Like other small landmasses surrounded by water that have emerged over different time frames—such as the Galápagos, Hawaiian, and Caribbean island archipelagos—New Zealand provides a dramatic example of how the temporal process of species diverging and adapting to their local environment can be seen from place to place. New Zealand is one of the oldest island chains. A dizzying number of endemic species of birds, making up 87 percent of the avifauna, have emerged on New Zealand. Of the thirty-two species of flightless birds, sixteen are now extinct. In addition to the small flightless wrens, the island chain is known for species such as the takahes (flightless rails), the Kakapo (a flightless parrot), and of course the kiwis. At one point, at least nine species of large, wingless ostrich-like birds called moas also inhabited these islands. But by AD 1400, just 250 years after the arrival of the first humans to New Zealand (the Maori), all nine species of moa had gone extinct, due to a combination of overhunting and habitat destruction. By the time Lyall set foot on the shores of Stephens Island, almost a third of New Zealand's unique species were already extinct due to Maori and European settlement, the habitats they destroyed, and the mammalian predators they brought with them.

Before Tibbles's arrival, there had never been a cat on Stephens Island. In fact, there had never been any mammalian predators on the island. Tibbles, along with her litter in utero, was the first to come ashore, early in 1894. A female cat can produce a litter of as many as eight kittens, sometimes more, and if a male is around, she can be impregnated again within days after giving birth. If an unrelated adult male is not around, siblings will eventually mate with one another, or offspring will mate with their mother. Once in estrus, cats will breed rapidly and often, and their populations will grow exponentially if left alone. Cats make the perfect pet for an isolated island inhabitant, in part because they can obtain most of their own food from their surroundings. Lizards, birds, or small mammals provide a sufficient diet. Cats are carnivores and need to consume primarily protein and some fat to stay healthy. They are ambush predators, sitting for long periods, motionless and quiet, waiting for the right time to pounce. They are quick and efficient and excel at what they do—otherwise they die. Cats have retractable, razor-sharp claws that extend from their strong paws to pin down prey. Once the prey is immobilized, cats inflict the kill bite with two sharp canines, usually to the neck, and quickly begin tearing into scales, fur, or feathers. Cats can kill animals as large as rabbits and squirrels, but their primary prey consists of smaller rodents like mice and voles as well as birds the size of (and including) sparrows and wrens. Cats do not always kill out of hunger. They seem to be stimulated by the chase and if not hungry will still kill; cat owners who allow their cat to roam freely may have received a "present" of a bird or mouse, a testament to their pet's predatory competence. Some scientists have postulated that the prey-return behavior serves as a way for an experienced cat to try to teach another cat or perhaps even a human to hunt. Others presume it is done by way of caching food or is perhaps an attempt to bring a once-playful toy to a safe place for use another time. Regardless, a single fed or unfed cat can bring home enormous numbers of furry and feathery presents to its owners. Tibbles, once on the island and allowed to roam, doing only what her instinct told her to do, soon began bringing an excited and curious Lyall small birds—probably sometimes whole and sometimes half-eaten.

No one knows what Tibbles was actually like as a companion, or whom she really belonged to. Like most cats, she probably had a fierce independent streak and was under the impression that the lighthouse keepers were there for her enjoyment and companionship rather than the reverse. Tibbles probably was not one to cuddle up on a lap or sleep around a head. As a kitten she was likely as silly and entertaining as any yarn ball–chasing kitten today. Once on the island and allowed to roam, Tibbles likely came and went at will. When not sleeping away long periods of time in the heat of the day, she explored the island, watching everything that moved, contemplating the chase of every twitch. Over time Tibbles probably became more and more wild. Certainly all her progeny were feral. Cats can "go wild" within a generation.

❖

Scientists do not know whether the Stephens Island Wren was at one time more widely distributed throughout New Zealand. It may be that through the destruction of habitats, combined with the spread of cats and rats, populations of this flightless bird had shriveled, and the isolated inhospitable Stephens Island served as a last refuge for the only remnant population. Fossil evidence points to this idea, but such evidence does not provide a definitive answer because it does not capture genetic or other differences that may have distinguished the Stephens Island Wren from other species. Equally plausible is the idea that fossil wrens found on other islands of New Zealand with physical features similar to the

Stephens Island Wren were actually different species. Given the biogeographic history of New Zealand, some populations of the Stephens Island Wren likely were isolated from other populations for millions of years, making this scenario quite plausible.

Typically, it is the combination of morphological and/or genetic differences that are used to delineate different species. The term *Biological Species Concept*, coined by evolutionary biologist Ernst Mayr, defines a species as a group of individuals that can potentially interbreed in nature. In the trenches of the biology field, this is seldom used as the only criterion for deciding when a population of organisms is a new or separate species. Taxonomists and systematists look at the color, patterns, size, and now the genes themselves to decide whether an animal might actually be a species new to science. These are the traits that develop largely through genetic mutations and/or long periods of reproductive isolation, such as might occur on islands or on the opposite sides of mountain ranges or rivers. Because whole specimens of similar wrens from other regions of New Zealand do not exist in collections, there is no way to know whether these other populations were actually the same species as those preserved from Stephens Island. Without this information, the true distribution of the wren, and all the possible causes of its demise in other New Zealand regions, if it was widely distributed, will never be known.

Nevertheless, a fundamental point is clear: By 1894 no one, including any of New Zealand's most renowned biologists, had recorded seeing the species, and Lyall perceived—having this wren in his hand on Stephens Island—that he was seeing something he had never seen before. Sitting down one evening next to the light that radiated off his paraffin lantern, an excited Lyall started to examine the most recent piles of birds brought home by Tibbles. Most were half-eaten, while others were almost completely intact. Lyall had been on the island only a short time, and thus far he could put a name to most of the specimens. Then he picked up the carcass of one peculiar bird. It was small, olive on the back, pale on the breast, with a scalloped brown fringe to the feathers. It had a narrow white streak above the eye, short wings, and a rather long, decurved brown bill. It reminded Lyall of the Rifleman

(*Acanthisitta chloris*), a similar species of small "wren" common in New Zealand and one he knew quite well. Lyall had likely seen specimen preparation on only a few occasions and had prepared a study skin himself only a couple of times. Nonetheless, he took his scalpel and made an incision along the small bird's reduced breast-bone straight down to the top of the belly. He worked his fingers under the skin, slowly pulling skin away from muscle. Eventually he worked his fingers in from either side until his fingers could meet. Using scissors, he snipped the bone just above where the rear end of the carcass attached to the tail and peeled the skin back from the body until he got to the wings. He could see that Tibbles had pierced the abdomen with her canines and had broken one of the wings, perhaps with the first swipe of her paw. He snipped both wing bones and cut the muscle. He continued to pull the carcass away from the skin, exposing the neck, and he quickly snipped that as well and removed the carcass from the skin. Carefully, Lyall peeled the skin over the skull until he could just see the edges of the eye sockets. He snipped a neat square of bone away from the back of the skull and then carefully scooped out the brain. Lyall knew he needed to remove as much tissue as possible from the specimen so it would dry quickly and to avoid a maggot infestation. He went back to work on the eyes and carefully snipped the thin layer of tissue surrounding the pupil, and pulled the eyes out of the skull. He reinverted the skin, which now had a small hole on each side where the eyes had been, back over the skull, and packed the rest of the skin with sheep's wool, re-creating the eyes, neck, and body. He then stitched up the incision and placed the specimen in the window to dry in the sun. Lyall repeated this process several times over the next few months, creating a series of at least fifteen spec-imens that eventually made their way to prominent ornithologists of the region and time, including Walter Rothschild, Walter Buller, and H. H. Travers (fig. 1.1).[1]

In just over a year, Tibbles and her offspring, their offspring, and all those that followed became wild and, according to Lyall, were "making sad havoc among all the birds."[2] Soon there were no wrens and few of the other species to be seen. It is not known ex-actly when this wren blinked off the earth for good. It could have been within a single year, but it was certainly not much more than a few years after Lyall and the other lighthouse keepers first made their way to Stephens Island that the wren disappeared. Lyall, his son, and perhaps a few others were likely the only humans to see the bird alive. On March 16, 1895, an editorial in the Christchurch newspaper *The Press* reported, "There is very good reason to believe that the bird is no longer to be found on the island, and, as it is not known to exist anywhere else, it has apparently become quite extinct. This is probably a record performance in the way of extermination."

A record performance that still stands today: extinct in perhaps a year and, ironically, at roughly the same time the identity of the species was first revealed to the world. A unique song, a lost language never recorded, and one now permanently silent. Fifteen specimens are all that is left of this species, and they exist in nine different museums around the world. Shortly after Lyall discovered the bird, his specimens had been bought, sold, and traded for amounts as high as $1,000 to $2,000 in current market valuation. The cats kept proliferating, and the fate of the birds of Stephens Island was clear. In 1899 the new lighthouse keeper was reported to have shot more than 100 feral cats in a period of ten months in an attempt to return the island to its pre-feline state. It took twenty-six years, but by 1925 the island was declared finally free of cats.

CHAPTER TWO

America's Dairy Land and
Its Killing Fields

> There are some who can live without wild things,
> and some who cannot.
>
> —Aldo Leopold

Almost 100 years after Lyall stepped onto Stephens Island, 8,500 miles to the north-northeast, Stanley Temple was wandering the islands of Wisconsin. Not islands surrounded by water and treacherous rocks but islands of grass (prairies, pastures, and hay fields) surrounded by row crops and forest. Temple, a professor of wildlife ecology at the University of Wisconsin–Madison, was studying birds in the quiet farm country of the rural Midwest with his graduate students. It was 1984, and Temple was interested in examining the wildlife benefits of federal and state grassland restoration projects, such as the U.S. Department of Agriculture's Conservation Reserve Program, that help farmers replace erodible croplands with permanent grass cover. The long-term intent of the USDA program was to create more grassland habitats within agricultural expanses, in order to help improve water quality, prevent soil erosion, and provide wildlife habitat. It was an early attempt to make farming more environmentally friendly, especially for birds and other grassland species that were losing their natural habitats as agriculture expanded and intensified. The wildlife benefits of the program, however, had not been thoroughly evaluated, and Temple, a pioneer in the field of conservation biology, knew that scientific evaluation was essential to avoid the unintended consequences that often result from such initiatives.

Wisconsin has a variety of natural habitats, including grasslands and savannas in the south and forests in the north. It is also known as America's dairy land and is heavily sprinkled with farms—dairy as well as those producing a variety of field crops. In 1984 approximately 70,000 farms covered the state—each with a farmhouse or two and multiple barns and other outbuildings. Temple started visiting some of these farms to set up his studies. As he roamed the state and its farmlands, he could not help but notice that most farms and associated grasslands were overrun with cats. Some farms were home to dozens of "barn cats" as well as free-ranging pets that roamed restored grasslands, preying on small rodents and birds in the precise habitats that farmers were being encouraged to manage for the benefit of native wildlife. Temple worried that by creating attractive nesting habitats close to farms and rural homes, restoration efforts might be luring birds to sites where they would be exposed to heavy predation—an ecological trap.

By this time, the domestic cat had been responsible for or had contributed to multiple animal extinctions on islands around the world, but little recognition or study had been given to its potential impacts in mainland areas. Temple decided to expand his research to address the specific question of how free-ranging cats were impacting Wisconsin's rural wildlife. He unwittingly began a scientific study that would create a firestorm, resulting in events that not even Temple, no stranger to controversy, could have predicted. A contentious debate began that would culminate in demonstrations and even death threats.

Tens of millions of people in the United States feel a deep bond and connection with cats. Cats are animals with fascinating and alluring personalities, but they can be destructive to native wildlife. Wild birds and mammals, however, also have rights that do not seem to receive as much attention as the claimed rights of cats to wander freely outdoors. These birds and mammals have become the victims in the brewing war between cat lovers and the people who support native wild animals. Stan Temple's study touched a dissonant chord. How should we deal with the animals that people have domesticated and enjoyed as beloved companions for thousands of years, but that when allowed to become feral or to freely range are capable of tearing away at the tapestry of life that has evolved since time immemorial?

🐾

Known as the Fertile Crescent, the area now composed of Iran, Iraq, Kuwait, Saudi Arabia, Bahrain, Turkey, and parts of Egypt, is considered the cradle of human civilization. Ten thousand years ago, it was the site of the first cultivation of agricultural crops such as wheat, barley, and lentils and the domestication of animals such as cows, goats, and pigs. Water was also beginning to be contained for drinking and irrigation there. The combination of carbohydrates, proteins, and water, along with their associated storage structures, allowed for complex human societies to evolve. As we see today in places like rural Wisconsin, such human developments attracted wildlife—including various species of granivorous mammals like the House Mouse (*Mus musculus*) and birds such as the House Sparrow (*Passer domesticus*)—that colonized structures and consumed grains. Known as peri-domesticated animals, because they thrive in close proximity to humans (*peri-* meaning "near"), these species can form new and complex food webs that sustain a diverse array of species, from plants to insects to mammals. It is thought that the surplus of mice and birds brought on by this convergence of events eventually led to the evolution of the domestic house cat somewhere in the Fertile Crescent. Exactly where this happened is not clear. However, cat skeletons have been found in close association with humans in ceremonial burial sites on the island of Cyprus, confirming that there was a close association with humans and cats as far back as 9,500 years ago.

Globally, there are forty native cat species in the family Felidae, indigenous to all continents except Australia and Antarctica. Most familiar are the large cats—Lion, Cheetah, Leopard, Jaguar,

Snow Leopard, Cougar (Puma), and Tiger. The remaining species, all smaller, are the African Golden Cat, Andean Mountain Cat, Chinese Mountain Cat, Asian Golden Cat, Bay Cat, Bobcat, Black-footed Cat, Canada Lynx, Caracal, Clouded Leopard, Eurasian Lynx, Fishing Cat, Flat-headed Cat, Geoffroy's Cat, Kodkod, Jaguarundi, Jungle Cat, Iberian Lynx, Leopard Cat, Marbled Cat, Margay, Pallas' Cat, Pampas Cat, Ocelot, Rusty-Spotted Cat, Sand Cat, Serval, Iriomote Cat, Oncilla, Colocolo, Pantanal Cat, and the Wildcat (*Felis silvestris*)—the progenitor of the newest and most controversial species of feline, the Domestic Cat (*Felis catus*). The Wildcat forms a complex group composed of at least twenty distinct subspecies including, for example: *Felis silvestris silvestris* (European Wildcat), *F. s. lybica* (Near Eastern Wildcat), *F. s. ornata* (Central Asian Wildcat), *F. s. cafra* (Southern African Wildcat), and *F. s. bieti* (Chinese Desert Cat). All are quite similar morphologically and genetically. The European Wildcat weighs between five and twelve pounds (depending in part on the sex), is gray with mackerel dark stripes, and, aside from being on the bulky side, has a face and body quite similar to any street-smart tabby (fig. 2.1). In fact, the European Wildcat is often mistaken for a feral domestic cat. Because of the European Wildcat's long history of interaction with feral cats, only a few of the remaining populations—hanging on in locations in Scotland, Switzerland, France, and Germany— are thought to remain genetically pure. In some places, such as northern and western Scotland, the number of pure wildcats that persist, if any at all, is unknown. The primary cause of the European Wildcat's decline was initially deforestation, but now inbreeding with feral domestic cats and contraction of the diseases they carry are considered the primary drivers. This is all somewhat ironic because this subspecies of the Wildcat, while genetically distinct, is considered one of the ancestors of today's domestic cat.

Recent genetic studies corroborate the notion that today's domestic cats evolved from several subspecies of wildcats and suggest that, of the five, the Near Eastern Wildcat (*F. s. lybica*) is likely the domestic cat's nearest relative. This also confirms the hypothesis that domestication of the cat occurred somewhere in the Fertile Crescent. Today, the domestic cat, largely due to selective breeding by humans and some genetic drift, has evolved into somewhere between forty and ninety different breeds. The count depends on the official cat registry that is consulted—for example, the registry of the Cat Fanciers' Association (CFA). A "breed" or "variety" is a group of animals within a species that share a suite of characteristics. If individuals from within a breed are bred together, including through selective breeding by humans, their progeny will retain the same characteristics because of the genetic underpinnings of their traits. Some of the cat breeds are as dissimilar as a Tiger and Lion and many look nothing like the ancestral Wildcat. They may be broadly classified as short-haired or long-haired breeds. Of the short-hairs, the varieties are as distinctive as the long-necked and ruddy-colored Abyssinian, the hairless Sphynx, and the smallest of the breeds, the Singapura. The long-haired breeds are equally varied and include curly-haired varieties such as the Selkirk Rex, the flat-faced Persians, and, the largest of all, the Maine Coon. Despite the impressive variation we see across the breeds of cat, it does not come close to what we see in dogs—160 to 400 breeds, again depending on the registry. Dog breeds include those as varied as Afghan Hound, Bulldog, Dachshund, Giant Schnauzer, Great Dane, Greyhound, Doberman Pinscher, Bernese Mountain Dog, and Chihuahua. The large variety in dog breeds has arisen because dogs have been bred by humans for much longer, perhaps even 1,000 years longer, than cats and for several purposes—to hunt, herd, and smell, to name a few. Cat breeding is a more recent phenomenon and has largely been done for cosmetic purposes by cat fanciers. Nevertheless, the initial domestication of the cat likely occurred simply through the accidental merging of several events.

In the way that Black Bears moved to foraging in garbage cans and White-tailed Deer to soft, lush (and seemingly the most desirable and expensive) plantings in people's yards, individuals from the various subspecies of the Wildcat probably went to locations where humans stored seeds, grains, and other food—locations that attracted rodents and birds. In this way cats and humans converged—in a relationship called commensalism—and this proximity, along with the fact that some individual cats were probably not as wild as others and thus were more tolerant of humans, led to the animal's domestication. Recent research now demonstrates that there may be a genetic basis for this "tolerant" or "tame" behavior. Cats with greater access to human-provided resources likely bred more successfully and, ultimately, produced a more human-tolerant breed of wildcat. Of course, it is equally plausible that early humans simply captured cats, alluring as they are, and tamed them into pets over many generations. No one really knows how it happened, and more than likely it was some combination of the above phenomena. What we do know is that descendants of these tamed cats spread, with human assistance, to almost every corner of the globe. They were tolerated by their human neighbors because of their supreme pet-like characteristics—soft fur, cuddliness, playfulness—and their uncanny ability to prey on and suppress the numbers of small animals considered pests by their owners.

When cats first arrived in New Zealand and Wisconsin, on thousands of islands in between, and elsewhere (except Antarctica) varies. But their initial spread was almost certainly linked to the movement and settlement of Europeans. Whether cats were brought as pets or mousers, or were simply stowaways on ships, is equally mysterious, although again, more than likely, it was a combination of all of the above. Edward II (1327–1377) has been cited as requiring all English vessels to have a cat on board for pest control, and this practice almost certainly contributed to the animal's global spread. The exact date cats arrived in the New World is not known, although they were apparently present by the second voyage of Columbus (1493–95). Cats were eaten by starving settlers in Virginia's Jamestown Colony in the early 1600s, and they are noted in historical accounts through the 1700s. Domestic cats have been in North America for at least 500 years, if not longer, and their spread, through both the unintentional and intentional actions of humans, earns them status as one of the most successful invasive species on earth. Their movement west across the United States, and to the state of Wisconsin, is another unknown. The first documented interaction between Europeans and North American Indians in Wisconsin came when Jean Nicolet arrived in 1634 to broker a peace deal between the Huron and Ho-Chunk nations. However, at that time, Native Americans kept dogs, and there is no record of cats making the trip. Trappers and traders continued to move throughout Wisconsin over the next 150 years, but it was not until the early 1800s that farming families from Europe and the eastern seaboard began homesteading the state and changing the landscape from prairies and savannas to pastures and croplands interspersed with farmhouses and woodlots. The arrival of *Felis catus* into Wisconsin from the east, either intentionally or unintentionally, likely occurred sometime during the initial European immigration into the area in the early 1800s. Within a few generations of their introduction, cats were there to stay.

❖

Stanley Temple is the Beers-Bascom Professor Emeritus in Conservation at the University of Wisconsin–Madison, as well as a senior fellow at the Aldo Leopold Foundation. He stands six feet tall, has a salt and pepper beard, wears glasses, and looks like a quintessential academic (fig. 2.2). Though Temple is soft-spoken by nature, his contributions to the field of conservation biology speak loudly. Temple has three degrees in ecology, a BS, an MS, and a PhD, all from Cornell University. His work has focused on the recovery of rare and endangered species. Temple spent thirty-two years as a professor at Wisconsin, in the same position held previously by none other than Aldo Leopold, who is considered the father of wildlife management. Leopold was the author of many scientific papers, popular articles, and, of course, *A Sand County Almanac*. Leopold's classic book, published in 1949, shortly after his death, was the culmination of his life's journey as a conservationist. It has sold millions of copies and has been translated into twelve different languages. Leopold decried the growing detachment between humans and all things wild and promoted the idea that our relationship with the natural world should be guided by ethical considerations. This notion motivated Temple as well.

Stanley Temple stepped humbly into Leopold's shoes, and over his academic and research career he has promoted Leopold's broad vision while focusing primarily on rare and declining birds. Temple and his students have helped save several bird species from extinction. He has written more than 320 papers and seven books, advised fifty-two master's candidates and twenty-three PhD students, and has won multiple professional and teaching awards. Quietly and thoughtfully, Temple has become one of the premier biologists in what was, at the start of his career, the emerging scientific field of conservation biology. His research on the biology of endangered species, his design and execution of recovery efforts, and his contributions to natural resource policy are some of his lasting contributions. His work with the Peregrine Falcon (*Falco peregrinus*), an endangered species that Temple studied early in his career, contributed to the species' rebound and eventual removal from the U.S. endangered species list. When he arrived on the island of Mauritius just a month after completing his PhD at Cornell, there were only seven remaining Mauritius Kestrels (*Falco punctatus*) in the wild—today there are 800. He helped launch decades of conservation work that helped save several species—including endemics—from extinction on the island. Halfway around the world, Temple and his students worked to save the endemic endangered Grenada Dove (*Leptotila wellsi*) on the island of Grenada and launched a recovery project for the species that included naming the dove the national bird. Closer to home, Temple and his students helped contribute to the development of techniques for the captive rearing, reintroduction, and eventual recovery of the California Condor (*Gymnogyps californianus*). Few living scientists have such a résumé.

Even as a child, Temple had a burning passion for watching and identifying birds and other wildlife. His mother, who did not share that passion, recognized her son's craving and allowed young Stan much freedom to explore nature on his own. She sent him on field trips with the Audubon Society of the District of Columbia to places like Hawk Mountain in Pennsylvania and the Maryland coast—both regional birding hot spots, especially during migration. As fate would have it, these birding trips were also attended by a scientist and nature writer named Rachel Carson, who took a special interest in the young Temple (fig. 2.3). Carson had already written an award-winning book, *The Sea Around Us* (which had been made into an Oscar-winning natural history movie in 1953), and likely was beginning research for what would be her signature contribution to the environmental movement, *Silent Spring*. Carson, like Leopold, had a deep love for nature, a desire for its preservation, and a commitment to communicating, through accessible and beautiful prose, why natural areas and the species they contain are essential to human life. Originally entitled *Man Against the Earth*, the book *Silent Spring* spoke to the dangers and impacts of the unregulated use of pesticides, specifically DDT (dichlorodiphenyltrichloroethane), on birds and other animals. Almost as important, this book was an indictment against the denial of scientific facts. Temple, by then a high school student working at the Cleveland Museum of Natural History, with a growing interest in birds of prey and conservation science, took notice when his inspiring mentor published *Silent Spring* in 1962. As a pesticide like DDT moves up a food chain—passed from plants to herbivores to predators of those herbivores—its concentrations tend to increase, through a process called biomagnification. Because birds of prey often consume herbivores and mid-level omnivores, they are prone to concentrate such pesticides, and by the 1960s they were becoming severely impacted by DDT. Populations of species such as the Bald Eagle, the Peregrine Falcon, and the Osprey were plummeting. Temple, influenced by two giants in the field of environmental science—Leopold, whom he never met but whose ethics profoundly influenced him, and Carson, who quietly and thoughtfully provided mentorship—was destined to tackle similar abuses against the earth in an effort to avoid the future silencing of our springs.

It took a long ten years after the publication of *Silent Spring* for DDT to be banned from use in parts of the United States. The delay was due largely to confusion manufactured by the deliberate misinformation disseminated by large corporate pesticide interests. Wisconsin was the first state to ban DDT's use. In June 1972 William Ruckelshaus, the first administrator of the newly created Environmental Protection Agency, announced the national banning of DDT, although it was still used sporadically around the United States until 1979. Like many other environmental contaminants, DDT still persists in nature, and to this day birds in some regions are still suffering from toxic levels of this pesticide.

Invasive species are, on one level, simply another form of an environmental contaminant; like DDT, they can cause great harm and, once introduced, can be exceptionally difficult to remove from the environment. The domesticated cat is one of the earliest invasive animal species on the planet, perhaps following closely behind only the House Mouse and the Black Rat (*Rattus rattus*). These two rodent species, indigenous to Asia, are transported by humans and considered cat prey. To be considered as an invasive species, the plant or animal, whose movement is assisted by humans, must be nonnative to a particular location and to spread like wildfire there, causing ecological damage to native species and the habitats they occupy. In some cases, invasive species can also impact human health—sometimes fatally—and wreak havoc on local and national economies. Think of Asian Tiger Mosquitoes (*Aedes albopictus*) carrying dengue and West Nile viruses, or of the common European Rabbit (*Oryctolagus cuniculus*) overgrazing massive areas of Australia's Outback and the South Island of New Zealand.

When an invasive species first settles and starts to reproduce and disperse in a new ecosystem, it needs to either occupy an available niche or outcompete a niche's previous inhabitant to be successful. Domestic cats have been successful on both counts—occupying a broad geographic reach and outcompeting previous niche inhabitants—while exponentially reproducing. In fact, it is this collective impact that now puts *Felis catus* on the list of the world's 100 worst invasive alien species. The animal's success as an invasive species was understood long before Temple started to study the impacts of rural cats in Wisconsin. In fact, the success had been illustrated unambiguously on islands (such as Stephens Island) and on wildlife—specifically birds.

Approximately 180,000 islands exist on earth. They vary in size and shape, in the amount of vegetation, and in their elevation.

They include continental islands, such as Greenland, Great Britain, and Madagascar—which lie on the shelf of a nearby continent—and oceanic islands, typically much smaller and of volcanic, tectonic, or coral origin (tropical islands). Because of their geographic isolation, islands share high levels of species endemism and biodiversity. Unfortunately, they also share high rates of species extinctions and/or declines in the population sizes of the endemic species they harbor. These rates are higher on smaller or midsize islands. Species with limited or no flying ability—like the Stephens Island Wren—are particularly vulnerable. Island species evolved with few to no predators, and as a result most have limited or no escape mechanisms. So, when a cat, rat, or mongoose—all extremely effective predators—makes it to an island, it is just a matter of time before significant population declines or extinctions of island fauna occur. To date, it is estimated, cats have been brought to about 10,000, or 5 percent, of the islands on earth.

In an article published in the journal *Global Change Biology*, Felix Medina and coauthors reviewed studies conducted on approximately 120 different islands around the globe of the impacts cats have had on endangered insular vertebrates. They concluded that cats have caused a population decline, a reduction in geographic distribution, or an extinction in 175 species of reptiles, birds, and mammals. Among reptiles, twenty-five species of iguanas, lizards, turtles, and snakes have been negatively impacted. In total, 123 species of birds have been negatively impacted, including songbirds, parrots, seabirds, and penguins. Among mammals, twenty-seven species—primarily rodents and marsupials but also a species of bat and even a lemur from the island of Madagascar—have been negatively impacted by cats. Overall, the domestic cat has contributed to or caused thirty-three (14 percent) of the 238 global reptile, bird, and mammal extinctions. Medina et al. concluded that this likely is an underestimate due to the lack of monitoring and research on most islands—especially in places known for high species endemism as well as a sizable presence of cats, including islands of Asia, Indonesia, Polynesia, and Micronesia. For comparison, there are no confirmed bird extinctions from the pesticide DDT.

Some of the thirty-three species whose extinctions are attributed to cats are still warm. Two of the twenty-two bird taxa include the Socorro Dove (*Zenaida graysoni*), last confirmed in the wild in 1972 (fig. 2.4), and the Hawaiian Crow (*Corvus hawaiiensis*), last encountered in 2002. Both species are different from the other thirty-one in that, although they are technically extinct in the wild, they are not fully extinct; individuals of both species were brought into captivity at various times and bred. Captive breeding, the removal of threats that drove these species to extinction in the wild, and the careful release and monitoring of individuals would need to happen if these species were to be successfully reestablished in the wild—approaches much like those Stan Temple took in his work with the Mauritius Kestrel and the California Condor. Each of these island extinctions is a story in itself, and cats were involved either directly or indirectly in the demise of each of the species in the wild. Retracing the steps of one such species' disappearance is helpful for those wishing to avoid similar disasters in the future.

🐾

Isla Socorro is one of the four Revillagigedo Islands lying off the southern tip of Baja California, just beyond the continental shelf in the Pacific Ocean. The archipelago is of volcanic origin, and its unique ecosystem supports at least sixteen endemic vertebrate taxa, all birds and reptiles. No native mammals occur on the islands. The island of Socorro has the richest biodiversity, including the aforementioned Socorro Dove as well as seven other species of endemic birds. Since 1867 there have been numerous expeditions to Socorro to survey its unique flora and fauna. One included Bayard Brattstrom, a UCLA PhD student who visited the islands in 1952 and published his findings four years later in the ornithological journal *The Condor*. One passage from his paper is almost prophetic:

> The future of the avifauna of the islands appears to be secure at present. There are no human inhabitants, and no mammals of any kind except the moderate and apparently stable population of sheep on Socorro. Few ships stop at any of the

Revillagigedo Islands, and the birds are seldom molested. The remote location and generally barren aspect of the archipelago have so far protected its biota from all except volcanic destruction. While this fortunate condition still exists, it may be hoped that the Mexican government will guard against the introduction of mammals such as rabbits, cats, goats and others that have invariably brought disaster to the flora and fauna of insular regions.[1]

Sadly, something did bring disaster to this small island after Bayard's 1956 observations. In March 1972 the state of Colima organized an expedition to Socorro Island to celebrate the 100th anniversary of Benito Juárez, a reform-minded politician who had served five terms as president of Mexico. The state of Colima was actually considering renaming the island in honor of Juárez. A published account of those festivities (Velasco-Murgía, 1982) provided the last scientifically credible sighting of the Socorro Dove in the wild. Velasco-Murgía noted that several visitors were seen killing the tame Socorro Doves with sticks, for no apparent reason. The Mexican ornithologist Juan Martínez-Gómez, who has spent most of his career studying the birds and their recovery on the Revillagigedo Islands, has since interviewed Mexican military personnel once stationed on the island, and although these are records by non-ornithologists, they suggest the Socorro Dove may have been seen as late as 1975. However, subsequent expeditions in 1978 and 1981 by Joe Jehl, Ken Parkes, and a returning Bayard Brattstrom failed to locate any individuals of the species. From what appeared to be a healthy population in the 1950s to the last survivors in the early to mid-1970s—the endemic doves of Socorro Island met a terrible fate.

Species extinctions are more often than not caused by multiple, cumulative, and interacting factors. While Jehl and Parkes found signs of feral cats throughout Socorro Island (cats were thought to have been brought to Socorro when a military installation was established in the late 1950s), and evidence implicated cats as preying on island endemics, subsequent research and interviews by Juan Martínez-Gómez failed to find convincing evidence that cats were on Socorro Island much earlier than 1970. Post-1972 Martínez-Gómez, while studying the endangered Socorro Mockingbird (*Mimus graysoni*), confirmed the presence of feral cats and their impacts. He identified body parts (feathers) of several endemic endangered bird species, including the Socorro Mockingbird and the Townsend's Shearwater (*Puffinus auricularis*) in cats' scat and stomachs. There is little question that feral cats contributed to the decline and eventual extinction of the Socorro Dove. But the degree of their responsibility remains unclear.

Martínez-Gómez postulates that the problem may have started in 1869 when sheep were first introduced to the island and eventually allowed to go feral. Their impact on the native vegetation, although not quantified, is indisputable. Over the next seventy-five years, visiting scientists, including Brattstrom, commented on the innocuous nature of the sheep population, but it was clear that the island's sensitive and critical habitat was undergoing a slow, cumulative change. Substantial overgrazing was triggering changes to the vegetation that would likely go unnoticed by the casual observer but certainly impacted birds like the Socorro Dove and Mockingbird, which were dependent on plants for food and cover. It is likely both species began to decline as soon as sheep were introduced. The decline accelerated just a few years after cats arrived. Because there were never any systematic monitoring schemes in place, it remains impossible to chart how quickly the decline actually happened. Socorro Doves were known to be particularly tame and were easy prey for cats and the occasional human with a battering stick. With an absence of cover, the dove had few places to hide, and remaining individuals were likely quickly dispatched by the growing feral cat population—much like what happened to the Stephens Island Wren. Thankfully, unlike the wren, Socorro Doves had been captured earlier in the twentieth century and bred in captivity. An eventual release back into the wild, once cats and sheep are finally removed from the environment, seems imminent. Whether the doves—like Stan Temple's Mauritius Kestrels—will be successful remains unclear.

Temple had been observing bird species flicker on and off the planet throughout his career. The specter of yet another invasive species wreaking havoc with birds was now before him in his home state of Wisconsin, as the number of free-ranging cats continued to increase. Temple was well aware of the strong evidence that implicated cats as a cause of species declines and extinctions on islands, and now he was seeing troubling numbers of these introduced predators in habitats surrounding farms and rural homes throughout Wisconsin. In many cases, these rural dwellings abutted grassland habitats containing species experiencing sharp declines, such as Eastern Meadowlark (*Sturnella magna*), Bobolink (*Dolichonyx oryzivorus*), and Henslow's Sparrow (*Ammodramus henslowii*). Temple knew it was time to collect data on the impact of cats on bird populations across the state and then, as a good conservation biologist, educate the public about the results, whatever the outcome. He secured funding from the USDA and the Wisconsin Department of Natural Resources, two government agencies that were promoting the creation of grassland habitats on farms to benefit declining wildlife.

Temple and his graduate student, John Coleman, set out to build a quantitative model, a necessary exercise if one hopes to understand some aspect of a complex system especially over a large spatial scale. Quantitative models are used in almost every aspect of science, in business, and even in everyday life; they simply vary in the complexity of their makeup and the uncertainty of their outputs. An example of a simple modeling problem might be to try to determine the cost of gasoline you would need to drive a car the length of the state of California. To determine this you need to know: (1) the average cost of gasoline, (2) the average miles per gallon for your specific make and brand of car, and (3) the length in miles of the state of California. Each of these estimates or "variables" has some degree of uncertainty. For example, the cost of gasoline will vary from station to station; the fuel efficiency will depend upon the speed driven, traffic, and the driver; and the exact miles driven will not be known until the end of the trip, making even this an estimate plus or minus some number of miles. The uncertainty can be estimated as well, so you know how much confidence you have in the final amount of money to put aside for your gasoline budget. In Temple and Coleman's case, the central question was straightforward—how many birds do cats kill each year across the rural areas of Wisconsin? The variables in the model were also fairly simple. First: how many free-ranging outdoor cats are there in rural Wisconsin; and second: how many bird kills per cat are there per year? Their model looked like this:

(number of rural cats) × (number of bird kills/rural cat/year)

To estimate the first variable—how many rural free-ranging cats are in the state of Wisconsin—they started off with the names and addresses of roughly 130,000 rural Wisconsin residents enrolled in the USDA's Agricultural Stabilization and Conservation Program. Almost all farmers and many non-farm rural homeowners were enrolled in this program, so this list ensured excellent coverage across the state. They then randomly selected 1 percent of potential respondents from this list, sampling from each of the seventy-two Wisconsin counties to ensure their results would be representative of all residents. A survey of twenty-two questions, along with a cover letter assuring anonymity, was then mailed to 1,324 Wisconsin residents on April 13, 1989. By November of that year, after sending reminder postcards and making a few phone calls, Temple and Coleman received about 807 usable responses—a 64.4 percent response rate. From these responses they determined that about 80 percent of respondents had between one and sixty cats (average of about five) and that the number of cats varied, depending on whether it was a farm or not and the type of farm. Livestock farms, for example, had up to nine cats on average. Temple and Coleman then took their data on cat density per type of rural residence and estimated the cat densities across the state. They came up with an estimate of 1.4 to 2 million free-ranging cats in rural Wisconsin. This equates to an average density of ten to fourteen cats per square kilometer (about 250 acres). These were essential estimates for developing the model and for better understanding the scale of the issue for bird populations. The researchers then had to determine how many kills were made per cat and how many of these were birds. This required capturing cats and collaring them with radio transmitters, so Temple and Coleman could measure several aspects of their behaviors.

Several studies had already been published estimating the number of kills an individual cat will make per year. Estimates indicated a range from zero to many more than a cat could possibly consume (one rural cat was recorded to have killed 1,690 animals over an eighteen-month period). Temple and Coleman used a variety of techniques to determine what cats were eating, including directly observing eleven radio-collared cats over 526 hours, analyzing the contents of 768 deposits of cat feces, and making 130 live-captured cats regurgitate their recently consumed prey with the use of harmless emetics. Cat owners participating in their study also reported 279 observations of their cats killing prey. Combined, these data revealed, as Temple and Coleman had suspected, that small birds and small mammals, such as rodents and rabbits, were typically the most common prey of free-ranging cats. Other studies using similar techniques have demonstrated that cats are generalized and opportunistic predators—they will kill anything that moves and is within their size range. Sometimes they will eat their prey and sometimes they will not—because of this variability Temple and Coleman considered it essential to combine as many types of studies as possible to develop their model estimates. Incorporating a literature review with their own data, they concluded that birds make up somewhere between 20 and 30 percent of the diet of free-ranging cats. The two researchers plugged the range of numbers from each of their components into the models and made their computations. They determined that rural free-ranging cats in Wisconsin kill, at a minimum, 7.8 million birds annually (at least 1.4 million cats times at least 5.6 birds killed per cat per year). Given that the densities of free-ranging cats in some areas of rural Wisconsin were several times higher than the typical combined densities of other midsize predators, such as foxes, skunks, opossums, and Raccoons, free-ranging cats clearly represented a major predator of native birds in rural Wisconsin. But what did these numbers mean?

Temple once again built a fairly simple model. He had detailed tracking data on cats, so he knew the typical size of a given animal's home range and the specific habitats where cats hunted. Bird censuses had been done in these habitats, so he also knew how many birds were present and available in each acre of habitat and within the average hunting range of a cat. What he concluded was that a minimum of 10 percent of small- to medium-size birds living within each cat's hunting range were typically lost to cat predation—a fairly significant take.

Most academic papers are read by peers at other universities and filed away to be periodically reviewed by other researchers exploring similar topics. But when Coleman and Temple's preliminary report was published in 1989, in the Fourth Eastern Wildlife Damage Control Conference program, and a longer report was published in 1993 in *Wildlife Society Bulletin*, the story reached the general public. Cats, it seems, are catnip for journalists. "I had no idea simply uncovering the facts would touch such a raw nerve with cat activists," Temple said many years later. "It was tough dealing with the nasty calls and ugly hate mail."[2] Temple even received death threats, after a newspaper article inaccurately implied that cats had been killed so their stomach contents could be analyzed. No cats were harmed in the course of the Wisconsin study, although some radio-collared cats did die from diseases and accidents. (Temple would receive more hate mail and death threats in 2005 when Wisconsin considered legislation that would essentially legalize the hunting of free-ranging cats—more on that later, in chapter 6.) All this uncivil discourse revealed a surprising disconnect: Many Wisconsinites (at least those who wrote letters to the editor and hate mail to Temple) were much more concerned that cats were being blamed for songbird deaths than with the fact that millions of songbirds were being killed. And some were more troubled about the possibility of cats being killed than they were about the life of a researcher.

Stan Temple actually likes cats and has owned several, which he keeps inside his rural Wisconsin home. He places value on his pets and knows they are safer indoors. But he also sees value in the lives of the songbirds and other wildlife species that are native to the land surrounding his home. More and more people are valuing birds and swelling the ranks of bird-watchers. Likewise, there are more cat owners in America now than at any time in history. But far fewer people, it seems, can summon affection for both cats and wildlife—and empathy for those they perceive to be on the "other side." As each side has swelled in numbers, the stage has been set for "bird people" and "cat people" to square off, forgetting, perhaps, that they are all animal lovers in the first place.

CHAPTER THREE

The Rise of Bird Lovers and Cat Lovers: The Perfect Storm

> A man who carries a cat by the tail learns something he can learn in no other way.
>
> —Mark Twain

On a sunny spring Saturday in 1919, an eleven-year-old boy named Roger Tory Peterson spotted a yellow bird in a park in Jamestown, New York, a thriving burg in the far southwestern corner of the Empire State. Set along the banks of the Chadakoin River, Jamestown was home to several woolen mills in the early days of its settlement. By the turn of the nineteenth century, it had become known as the Furniture Capital of the World—thanks to abundant timber and the presence of many Swedish immigrants who arrived with fine craftsmanship skills. Though the forests of oak, maple, and white pine surrounding Jamestown were aggressively logged to satisfy the appetites of the furniture factories along the Chadakoin's banks, the woods of the Hundred Acre Lot—a neighborhood park on the south side of Jamestown—were spared.

That Saturday morning—a morning that many herald as the birth of bird-watching in America—young Roger accompanied his seventh-grade teacher, Blanche Hornbeck, and some fellow students on a field trip to the Hundred Acre Lot. Miss Hornbeck had formed a Junior Audubon Club, whose members met after school to study Audubon Society educational leaflets and try their hands at copying bird illustrations from E. H. Eaton's two-volume *Birds of New York State*. The Hundred Acre Lot had recently been purchased by the city of Jamestown for $10, in part through pennies raised from students. Strolling through the park, Roger and a friend came upon a bird they identified as a flicker (now known as the Northern Flicker, *Colaptes auratus*) on the trunk of a tree, an experience that would prove life changing. Peterson would later write:

> Its head was tucked under its wing coverts. It was probably exhausted from migration, but we thought it was dead. We stood and stared at it for a while, examining its beautiful plumage. When I reached out to touch its back it exploded with life—a stunning sight, flying away with its golden underwings and the red crescent on its nape—I can see it now—the way it was transformed from what we thought was death into intense life. I was tremendously excited with the feeling which I have carried ever since, of the intensity of a bird's life, and its apparent freedom, with this wonderful ability to fly.[1]

A perhaps equally propitious moment: during one afterschool session of copying bird illustrations, Peterson painted a watercolor of a Blue Jay; he later would recall that from that moment on, he wanted to be a bird painter.

By all accounts, Peterson's youth was not easy. There was never quite enough money for his family to make ends meet, and his father, a first generation Swede who had begun working at age ten, had limited patience for his son's burgeoning naturalist impulses. How would watching birds and collecting butterflies ever put food on the table? Perhaps it was a nod to his father's pragmatic world view that brought Roger to the National Furniture Company to paint not birds, but Chinese-inspired designs on lacquered cabinets. Fortunately, one of his managers saw the young man's promise and encouraged him to attend art school. In 1927 Peterson left Jamestown for New York City, where he attended classes at the Art Students League and eventually the National Academy of

Design. While polishing his drawing and painting skills (and still decorating furniture to pay his bills), Peterson was also able to spend time at the American Museum of Natural History, where he began to receive his first formal ornithological training. In 1931 he moved to the Boston area, where he took on a teaching position at the Rivers School, a prestigious boys' academy. This provided his first opportunity to share his love of bird-watching with students. A dedicated teacher by day, Peterson made time to paint and write about birds—a passion that consumed him (fig. 3.1). His off-hours efforts resulted in a first book, *A Field Guide to the Birds*, which would have such an impact it would change the way people experienced the natural world. When the richly illustrated manuscript made the rounds with a literary agent, four publishing houses turned it down. But Boston-based Houghton Mifflin took a gamble after Peterson visited in person with his work. The initial print run in 1934 of 2,000 copies sold out in less than a month—no small feat for a book on birds during the depths of the Great Depression. (The Peterson Field Guides series has gone on to sell more than 7 million books, many of which are still in print today.) In his unassuming way, Roger Tory Peterson had created a whole new category of nature book—the field guide—a genre that would democratize birding for Americans and set the stage for the ecological awakening in the mid-twentieth century buttressed by the likes of Aldo Leopold and Rachel Carson.

🐾

Just as Peterson introduced innovations in field guides that benefited countless bird-watchers, a mixed-breed female tabby cat in Arizona named Tardar Sauce helped blaze new trails of Internet inanity, confirming YouTube as the greatest place to waste time online. Tardar Sauce is better known as "Grumpy Cat" (fig. 3.2). The edges of the cat's mouth are pulled forever downward, giving the animal its eponymous dour expression, which is actually believed to be the result of feline dwarfism and an underbite. Since her online debut in 2012, Tardar Sauce—with a bit of help from her owner and a manager—has catapulted into the mainstream, earning front-page and cover stories in such publications as the *Wall Street Journal* and *New York* magazine and a slot with Anderson Cooper, not to mention in excess of 8.5 million "likes" on Facebook. What did Grumpy Cat do to gather such acclaim? In one video clip she lies on her back. In another she lies on her stomach. In a third she yawns. Thus far Grumpy and her entourage have taken her notoriety beyond various social-media properties and parlayed her trademark frown into a cable-TV movie, two best-selling books, and an iced-coffee beverage named, of course, Grumppuccino.

Grumpy's success underscores the fact that, as a people, Americans have a tremendous love and fascination for cats. The only thing we may love more than cats is online memes that highlight our love of cats. Not that long ago, however, cats were much less likely to be near our hearts, let alone inside our homes. As noted in chapter 2, the domestic cat's relationship with humans has evolved over 10,000 years, a long and not always amiable history. We have loved and worshiped cats, hated and condemned them. We have used them as pest-control experts and have placed them near the center of our lives. The cat's high status in ancient Egyptian culture is well documented. Egyptian cats likely earned human esteem in the same way their Victorian counterparts did: by reducing the vermin populations around granaries and vanquishing the occasional cobra. Records suggest that killing a cat was an offense punishable by death. The goddess Bastet, who was often represented as a domestic cat, was worshiped as a symbol of fertility; at festivals celebrated in her name, hundreds of thousands of cats were sacrificed, mummified, and buried. The idea of cats as figures of feminine lust has persisted, as a listen to the Rolling Stones' "Stray Cat Blues" will attest.

In Europe by the Middle Ages, the status of cats had taken a decided turn for the worse. Closely associated with witches and Satan (thanks to their nocturnal inclinations), felines were hunted mercilessly. In the fifteenth century, the Belgian town of Ypres took the revulsion of cats to a new level, mounting a feline festival— more an anti-festival—that culminated with the town jester flinging live animals from a belfry. This festival continued until 1817.

By the dawn of the twentieth century, cats were no longer considered agents of evil. But in the United States at the time, their role was that of a worker, rather than a companion. Farmers and other rural dwellers kept felines around to keep down populations of mice, rats, and other vermin. (It must be remembered that in 1900 a majority of Americans still lived in rural areas.) They were unlikely to set up a kitty bed for Tabby by the stove or in the corner of the bedroom; she bedded down in the barn or maybe in a box on the porch and, beyond the occasional saucer of milk or invitation to sit by the hearth, fended largely for herself. But social changes were coming that would bring cats inside and entwine them more and more tightly around our lives.

The years surrounding the beginning of the twentieth century marked a human exodus from rural settings and small towns to America's cities, as tens of millions were drawn by the availability of industrial jobs. Most urban immigrants found themselves living in cramped apartments or tenements; there was hardly enough room for a family, let alone pets, especially as the need for controlling varmints was less pressing in a decidedly nonagricultural setting, though the tenements certainly had no shortage of rodent inhabitants. Several practical considerations beyond space impeded the adoption of domestic cats into urban homes for several decades: First, there was food. Cats need a high-protein diet, and without the ability to access the outdoors and hunt, they would come up short on meat; few families at the time could afford additional protein sources for their animals. Lack of outdoor access created a second problem. Assuming your kitty could get enough to eat, where was she going to defecate and urinate? Kitty litter had yet to be created, and very few urbanites would be willing to endure the mess and stink of cat feces and urine in tight quarters. The threat of too many cats was a third obstacle to apartment-dwelling cats. Practical spaying and neutering techniques for pets did not come along until the 1930s. Apartment dwellers with a female cat in heat found themselves either with a very unhappy animal or, if they kept a male cat (or another male was in the building and not locked away), with a boatload of cats. Cats average three litters a year; the average number of kittens in each litter is four to six. Kittens can come into estrus as early as four months after being born, so the numbers of cats can multiply *very* quickly!

This is not to say that cats were strangers to America's cities. Unlike dogs, which were often kept as pets earlier in the twentieth century, cats had a more ambiguous position in urban households; Katherine Grier described them as independent contractors in her informative book *Pets in America: A History*. They were prized as ratcatchers, especially around urban stables, where there was feed (and hence, vermin), but generally not full-time members of the household for the reasons highlighted above. There were exceptions, of course; Mark Twain loved cats and was sometimes spied walking about with one of his cats—named Lazy—draped around his neck like a stole. As public concern shifted to order and cleanliness, the numbers of "tramp cats" in cities became problematic. Free-ranging cats were seen as carriers of disease and hence a public-health threat and were killed off in large numbers. Concern over such draconian methods led some citizens to seek more gentle means of control. The Animal Rescue League of Boston and the Bide-a-Wee Home in New York championed adoption programs for stray cats and dogs, as well as more humane methods of euthanasia.

🐾

Killing animals was not on Roger Tory Peterson's agenda as he set out to create his field guides. In this respect, he differed significantly from his literary forebear John James Audubon, whose *Birds of America* was released over eleven years, from 1827 to 1838. Though both books shared the topic of American birds, the similarities stop there. *The Birds of America* is not exactly compact; the original printing measured more than three feet tall by over two feet wide! Nor was it economically priced; the first set cost $870 in 1830's U.S. dollars (that would be close to $60,000 in 2015 U.S. dollars). Conversely, the 1934 edition of the very portable *A Field Guide to the Birds* retailed for $2.75 (fig. 3.3). But it is in its content that Peterson's work most dramatically differed from that of Audubon and other predecessors. These earlier bird portraitists created beautiful, even resplendent likenesses of the birds by "posing" birds they had shot, using wires and other apparatus. Indeed, for many enthusiasts up until the early twentieth century, birding could be equated with hunting. Elliot Coues, an eminent American ornithologist of the 1870s, provided the following advice to prospective bird-watchers: "The double barreled shotgun is your main reliance. Get the best one you can afford for your particular purpose which is the destruction of small birds with the least possible damage to their plumage. Begin by shooting every bird you can."[2]

Peterson's paintings were competent enough, but what made them special was the manner in which they highlighted the telltale markings of each species—its "field marks"—so it could be easily differentiated from similar species and identified from a distance. The field marks—say the *pink bill*, the *white patch on front of face* and the *variable black bars on belly* of the Greater White-fronted Goose—were called out with arrows. Instead of relying upon a firearm to connect with birds, amateur naturalists could carry a Peterson field guide and a pair of binoculars. They became observers, rather than hunters. Bird-watching with binoculars could be done almost anywhere and anytime.

The inspiration for Peterson's breakthrough idea did not come from the New York art establishment or the Boston bluebloods of the Nuttall Ornithological Club, of which Peterson was a member (at the time one of the country's most elite birding groups). Instead, it came from a book that he had devoured as a young boy, *Two Little Savages*, by Ernest Thompson Seton. Written in 1903, the book chronicled the adventures of Yan and Sam, two Canadian boys, who would steal away to the woods as often as their schedules would permit and imagine how the native peoples would have lived. Peterson identified with Yan. In a chapter entitled "How Yan Knew the Ducks from Afar," the protagonist identified a prominent shortcoming of the birding literature available at the turn of the century—and, for that matter, in 1933:

He found lots of difficulties and no one to help him, but he kept on and on . . . and made notes, and when he learned anything new he froze on to it like grim death. By and by he got a book that was some help, but not much. It told about some of the birds as if you had them in your hand. But this heroic youth only saw them at a distance and was struck. One day

he saw a wild Duck on a pond so far away he could only see
some spots of colour, but he made a sketch of it, and later he
found out from that rough sketch that it was a Whistler, and
then this wonderful boy had an idea. All the Ducks are different; all have little blots and streaks that are their labels, or
like the uniform of soldiers. "Now, if I can put their uniforms
down on paper I'll know the Ducks as soon as I see them on
a pond a long way off."[3]

Peterson frequently acknowledged the impact of Yan's approach
to bird identification; in fact he cited it in the introductions to the
1934, 1939, and 1947 editions of his *Field Guide to the Birds*.

"When Roger Tory Peterson showed people what to look for
to identify a Cape May Warbler, he made bird watching something that the general populous could do," Bill Thompson, editor
and publisher of *Bird Watcher's Digest*, shared. "You didn't need
a shotgun anymore. Just about anyone could get their hands on a
Peterson guide. With the field guide, you weren't looking at just a
bird anymore. It was a *Blue Jay*. The field guide also made it possible for a large community of bird-watchers to develop. Like the
telephone or rural Internet, it connected people."[4]

Noble Proctor, the 2013 recipient of the American Birding Association's Roger Tory Peterson Award for Promoting the Cause of
Birding, believed that Peterson's field guides are responsible for the
rise of the green movement:

> By making Americans aware of birds, the field guides opened
> up our environmental thinking. There were more types of
> birds than crows, robins and "little brown ones." The guide
> encouraged a closer look by rewarding it. They turned aware-
> ness into knowledge. Before you can go out and save any-
> thing, you have to know what it is. Knowing the names of
> things became an essential step toward involvement.[5]

Every movement needs a manifesto, and *A Field Guide to the
Birds* was the catalyst that propelled American bird-watching
forward. A number of other factors coalesced to transform and

popularize the pastime of birding. One was the increasing availability of quality optical equipment. At the time of the release of
the first field guide, many birders were using field glasses that were
castoffs from military personnel in the family. Advances in optics
technology—some acquired from Germany during the occupation
after the war—put better binoculars and sighting scopes in birders' hands at a much more affordable price. Larger cultural and
economic factors that came to bear on post–World War II America also abetted our growing passion for birds. With automobiles
more accessible to the working class than ever before, American
families flocked away from the cities to the suburbs. In the suburbs, people had yards, where they could toss a football, barbecue
hamburgers, and hang bird feeders. Thanks to the general pros-
perity and increase in standards of living that many Americans
experienced in the 1950s, people of more modest means had more
leisure time at their disposal than ever before—time to pursue hob-
bies like birding. And they had the automobiles to transport them
farther and farther afield.

Why, the non-birder may ask, all the fuss? What's the appeal of
birds? Bill Thompson has contemplated that very question:

> Is it their dazzling plumages? Is it their ethereal, musical
> songs? How about their courtship rituals and devotion to
> mate and offspring? These are all admirable explanations
> for our love of birds, but I think it's even more simple than
> that. After all, birds have been doing something for eons that
> humans have only figured out in the past 100 years—flying.
> It's the freedom, the power of the ability to fly that draws us
> most to birds. They are not bound, as we are, to the ground.
> Birds ignore the bonds of gravity, and we wish that we could,
> too. And so we watch birds in wonder and seek them out
> wherever they occur.[6]

Flight allows some species to cover thousands of miles every year
during their biannual migrations. Perhaps it is these migrations that
mesmerize and capture the imaginations of millions of dedicated
bird-watchers. Migrations can span hemispheres, some species

flying almost pole to pole, sometimes in flocks of tens of thousands of individuals, and some flying only at night to avoid predators. And despite such long-distance flights, every year they can return to within a few hundred yards of the same wintering and breeding territory they used the previous year. Billions of individual living torpedoes of feathers move every spring and fall—through the air— all over the world. These animals connect our cultures, our countries, and our continents in ways we do not completely understand. They move from tropical jungles to temperate and boreal forests to grasslands, deserts, marshes, and backyards. These remarkable voyagers, whether hawks, rails, shorebirds, or songbirds, are the ultimate travelers, seemingly always on the go, with no luggage, moving only to survive, stopping occasionally to refuel, fleetingly captured by the scanning binoculars and straining necks of bird-watchers seeking to experience this mysterious behavior.

As exotic as birds can be, it is their commonness that makes our connection with them so easy. Observing birds, in our yards or in faraway lands, connects us to wild nature—a connection that is increasingly frayed in modern society. Richard Louv coined the phrase "nature-deficit disorder" to describe Americans'— especially children's—growing disengagement with the natural world and the resulting range of behavioral problems, including attention disorders, anxiety, and depression. Bird-watching engages one's senses and one's intellect in the act of identification; you must concentrate, look, and listen, and then process the stimuli gathered to understand what bird you are observing. In a way, every identification poses a challenge, a puzzle, a problem to solve. Every new species encountered promotes a hypothesis-testing exercise. Birding also satisfies the very human desire to gather and save—and to compete to see who can gather the most. It can be a game. In the late 1800s this gathering took the form of shooting and stuffing specimens. Today it finds its expression in the life list, the race to announce on Twitter the sighting and location of a rare species or to compile the largest list on eBird, a novel online checklist program for the birding community.

Birders have often been dismissed as eccentric, if not slightly daft, by the popular culture; fans of *The Beverly Hillbillies* may recall the character Miss Jane Hathaway, ever clad in tweeds. Even the cast of Steve Martin, Jack Black, and Owen Wilson could not make birding seem hip in the 2011 comedy *The Big Year*; perhaps the pastime simply does not translate well to the screen. Are birders truly unconventional outcasts, fringe characters not to be taken seriously? A 2011 survey conducted by the United States Fish and Wildlife Service (USFWS) on the economic impact of bird-watching revealed that whether they are serious birders or casual bird-watchers, they are much more mainstream than most would think. In fact, they probably are your next-door neighbors, or at least those living just a few doors down. The survey concluded that there are some 47 million active birders over the age of sixteen in the United States. The USFWS identifies birders as individuals who have either closely observed or tried to identify birds around the home or taken a trip one mile or more from home for the primary purpose of observing birds. Who are all these millions of bird enthusiasts? Eighty-eight percent—or 41 million—are backyard bird-watchers, folks who might hang a few feeders or a chunk of suet outside and keep a field guide on the breakfast nook table. The remaining 6 million, plus 12 million of the casual birders, take trips away from home to view birds. Birders skew a little older in age and tend to have a higher income and a higher level of education than the average American, are slightly more likely to be female, and are much more likely to be white. Rural residents are more likely to be birders than urban residents (on a per capita basis). People living in southern states are more likely to be birders than their fellow Americans in the Midwest, the Northeast, and the West.

A number of groups provide a sense of community for America's birders and bird-watchers. There are the American Birding Association, with approximately 12,000 members; the American Bird Conservancy, 10,000 members; the Cornell Laboratory of Ornithology, with about 70,000 members; and the National Audubon Society, with over 464 state chapters and 450,000 members across the United States. Then there are scores of other state-specific Audubon organizations not affiliated with National Audubon. National Audubon's activities go far beyond organizing bird-watching outings. Members partake in grassroots activism and restoration activities and conduct "citizen science," most notably through the annual Christmas Bird Count. The organization also funds habitat protection, employs scientists and lobbyists to shape and institute conservation plans, and strives to educate the population through a variety of programs ranging from Audubon centers to *Audubon* magazine. National Audubon spent close to $74 million in 2014 to support conservation efforts, and garnered protection for almost 120 million acres of habitat.

The pastime of bird-watching has emerged as an economic as well as a political force. In 2011 American birders spent an estimated $15 billion on trips (food, lodging, transportation) and $26 billion on equipment (binoculars, cameras, camping gear). These expenditures amounted to a total industry output—the so-called ripple effect on the economy—of $107 billion.

🐾

As an economic engine, cats are no slouch either. Just as new products and technologies like Peterson's field guides and improved optics helped foster a culture of bird-watchers in America, modern developments such as canned pet food (and refrigerators to keep it fresh), as well as advances in other food products, made it easier and more cost-effective for people of modest means to take a cat into their home. The "invention" of kitty litter in the late 1940s— basically Fuller's earth, a type of clay that proved perfectly suited for soaking up the ammonia scent of cat pee—made keeping cats inside a much more palatable venture. Improvements in spaying and neutering procedures, prompted in part by the veterinary industry's increasing focus on smaller animals, also made keeping cats inside a more pleasing and manageable prospect.

Today, well into the twenty-first century, cat ownership has reached an all-time high in the United States, and an estimated 90 million pet cats live in up to 46 million American homes. According to the Pet Food Institute, cat food sales in the United States increased from $4.2 billion to $6.7 billion between 2000 and 2013, an increase of over 50 percent. Societal trends have contributed to the explosion of cat ownership. We are now a much more urban society, with 80.7 percent of the population (more than 249 million people) living in or near cities. This has distanced us from wild animals and their natural world—bringing pets into our homes satisfies the inner need of people to connect to animals. The United States has also seen an increase in one-person households, from 17 percent of all households in 1970 to 27 percent (nearly 32 million individuals) in 2012. Cats provide companionship for people living alone, without the need for walking or pooper-scooping.

"I believe that part of the allure of cats as companion animals is the fact that they have a bit of wildness in them," mused Sharon Harmon, president and CEO of the Oregon Humane Society. "Cats aren't quite tame. They're only one generation away from being in a wild state. We embrace that wildness."[7] Cat owners are happy to elaborate on why they like to have cats in their lives, as a cursory review of the website Catster revealed:

- They are better alarms than any snooze button or clock will ever be.
- They give you someone to talk to. Sometimes they even listen.
- They can eat the same thing every day and not complain.
- You can talk like a baby and not feel silly.
- You don't have to take them outside to do their business.
- They know when you need a little extra cuddle.
- They can stay home alone all day and the house will be fine.
- They can turn almost anything into a toy.
- They love you unconditionally . . . as long as you feed them.
- How they fly through the air and twist and do crazy acrobatic moves to catch a toy.
- Cats are the best work distraction ever.
- And of course . . . laser pointers.[8]

There's also evidence that cats contribute to the well-being, both physical and psychological, of their owners. Feline company has been shown to lower blood pressure and improve the moods of some people. Pets (cats and dogs) have been used therapeutically for children with disabilities (most notably through a program called Pets As Therapy, or PAT). It is heartwarming to watch the faces of children who are unwilling to talk to or be touched by people light up around cats, both speaking to their feline companions and stroking their soft fur.

While "owned" cats like Tardar Sauce—and even others with fewer than 8.5 million likes—lead a fairly cosseted life indoors, the estimated 60 to 100 million cats in America that are "unowned" and live outside endure a bleak existence that puts them on a collision course with our wild birds, one that seldom ends well for the birds. Unowned cats without veterinary care are prone to disease (including feline leukemia, renal failure, feline panleukopenia, plague, rabies, and toxoplasmosis, as we will explore later). They are vulnerable to predation by other animals, especially Coyotes and, to a lesser extent, eagles, owls, foxes, and Raccoons. And they are frequently hit by cars—the most common cause of demise in outside cats. Such are the hazards if they survive to adulthood, but estimates suggest that 50 to 75 percent of kittens born outdoors do not, dying from exposure, parasites, and disease. If they do reach adulthood, the life expectancy of an outside cat without caregivers providing regular feeding, water, and sometimes makeshift shelter is two years. Outside cats that receive such care have a much longer life span, averaging ten years. The average life span of an inside cat is thirteen to seventeen years, depending on the breed.

It is not possible to monitor the general *happiness* of an outside cat, but we do have a general idea of these animals' habits. Some—especially truly feral animals—are unsocialized and reject any interaction with humans. Others—strays that have been lost or abandoned—will seek out human contact. Outside cats that rely on humans in part for their sustenance tend to congregate in groups or colonies. Such colonies are often anchored around female kinship; males, as well as animals that do not rely on human assistance, tend to be more solitary. The day-to-day activities among outside cats are largely dictated by the availability of food. On the Japanese island of Ainoshima, for example, where cats had access to reliable food sources from the island's refuse pits, animals were observed to rest up to nineteen hours a day. In South

Australia, where the climate is quite arid and food is scarce, some free-ranging cats were constantly on the move, ranging up to fifty square miles. Outside cats in an American suburb might "make the rounds" from porch to porch where food has been provided by interested humans, grabbing a finch or field mouse as opportunities arise. Like their wild brethren, outside cats are most active in lowlight hours, particularly at dusk and dawn. This is especially true of outside felines living in urban or suburban settings, as the animals may prefer to avoid human confrontations.

Our relationship with outside cats has always had its complexities. Today it even extends to how we label them. A white paper by the Humane Society of the United States (HSUS) points out that in the scientific literature on the topic there are more than thirty terms used to describe outside cats, ranging from "feral" and "invasive" to "pet" and "house cat." How we define them is influenced by sociological as well as biological constructs: what their ownership status is as well as where they spend their time. The term "feral" is often used as a catchall for outside cats, though it technically should apply only to animals that have completely returned to the wild, have no reliance upon humans for any sustenance or shelter, and reject any interaction with humans. Other descriptors for outside cats encountered in an urban or suburban environment include "semi-owned," "street," "stray," "colony," and "neighborhood," all of which imply a level of human dependence and thus are more accurate. The waters are further muddied by the fact that many "house" or "pet" or "owned" cats are allowed to wander outside, some for as long they wish. The International Companion Animal Management Coalition recognizes three categories of human-cat relationships: owned, semi-owned, and un-owned. For the purposes of this book, the term "free-ranging" seems most appropriate to refer to any cat that spends at least some time outdoors, as it describes the animal's ability to move about as it wishes, without any assumptions about its interaction—or lack thereof—with humans.

Whatever you call free-ranging cats, their number is growing in America, and cat abandonment is a significant contributing factor. Abandonment is defined in the state of New York under

Statute 355 of the New York Agriculture and Markets Law, for example, as follows:

> A person being the owner or possessor, or having charge or custody of an animal, who abandons such animal, or leaves it to die in a street, road or public place, or who allows such animal, if it becomes disabled, to lie in a public street, road or public place more than three hours after he receives notice that it is left disabled, is guilty of a misdemeanor, punishable by imprisonment for not more than one year, or by a fine of not more than one thousand dollars, or by both.[9]

There is no way to accurately assess the number of animal abandonments—people do not tend to boast about such insidious (and in many jurisdictions criminal) behavior—but the American Society for the Prevention of Cruelty to Animals (ASPCA) reports that 6 to 8 million companion animals (mostly cats and dogs) enter shelters nationwide every year. They are released in parks, at highway rest stops, on university campuses and military bases, or are simply left behind in apartments or houses as their owners move on. According to a study conducted by the National Council on Pet Population and Policy, cats are abandoned for a variety of reasons, including there being too many in a house; allergies; moving; cost of pet maintenance; landlord issues; failure to find homes for a litter; house soiling; personal problems; inadequate facilities; and a cat that does not get along with other pets. Most abandoners probably think that cats can fend for themselves. Most cannot. Because of the nature of the act of abandonment—often under the cover of night or in an isolated place—and the fact that cats are unable to testify on their own behalf, most offenses go unpunished. In his short story "The Good Work of Chickens," Richard Brautigan conceived an intriguing retribution for the cruelty of abandoners. After witnessing a dog being deserted by its owners at a rest stop, the story's narrator traces the driver's license plate and commissions a dump truck to deliver a ton of chicken shit to the deserter's front porch.

Free-ranging cats arouse strong and conflicting responses in the people they move among. Some focus on the animals' more disruptive behaviors, such as digging, defecating, and spraying around non-owners' homes, meowing and fighting when females are in estrus, and preying upon birds drawn to bird feeders.

But others take a passionate and committed interest in free-ranging cats. These people—sometimes referred to as colony caretakers—provide food and water for the animals, occasionally shelter them, and will sometimes facilitate veterinary care for the cats (including spaying and neutering), often paying for a vet's services out of their own pockets. Some care takes place in an ad hoc manner, with people occasionally placing a dish of table scraps or cat food in the yard of their apartment complex or out behind their shed. Others act in accordance with the strictures of a more formalized regimen of care, dictated by one of the hundreds of nonprofit organizations that have sprung up in recent years to assist free-ranging cats. Regardless of the level and type of care provided, most caretakers act out of genuine compassion for sentient creatures that they feel have received a raw deal in this life.

One such nonprofit, Alley Cat Allies, bills itself as "the only national advocacy organization dedicated to the protection and humane treatment of cats." There are other such organizations, but Alley Cat Allies is certainly one of the most vocal and the best funded. Among its activities, the organization advocates for pounds and shelters to keep public records of animal intakes and kill rates (for greater accountability); mobilizes and educates the public to end the killing of cats and to protect and improve their lives; and provides a clearinghouse of information for caregivers of free-ranging animals. Alley Cat Allies is not an isolated grassroots organization; it boasts a network of 500,000 supporters and is a loud and frequently strident voice on issues concerning the welfare of free-ranging cats, with an estimated annual budget of $5 million. Its members are committed to and passionate in their beliefs. Regrettably, many of their efforts fail to recognize at least two realities: enabling cats to roam freely outside (1) shortens the lives of a vast number of birds, amphibians, reptiles, small mammals, as well as the cats themselves, and (2) allows outdoor cats to spread disease that impacts not only wildlife but also humans. These points are discussed in depth in chapters 4 and 5.

It is not easy to find cat caretakers who will speak about their endeavors or introduce outsiders to their colonies. Organizations are loath to share members' identities, perhaps wary of an "outing" of sorts or some kind of retribution. Caretakers themselves, if identified, often are an equally furtive lot; one has the sense that they fear outsiders will attempt to harm the members of their colony, thus it is best to keep their location a secret. One organization based in Toronto, Animal Equity, has created profiles of several colony caretakers in the region. These portrayals, from the organization's website, shed light on caretakers and their motivations:

Robin S.

Robin is well known among cat rescuers, feral cat colony caretakers and cat lovers in Toronto. For the last 7 years, Robin has rescued many feral/stray cats and brought them to safe and warm forever homes. Cat colony maintenance requires a lot more commitment than most would imagine. Robin says that every day, including Christmas, she visits her colonies to feed the feral cats who depend solely on her for their survival. Sometimes colony caretakers face harassment by people who do not like feral cats in their neighborhood. To minimize conflict, Robin waits until the cats finish their meals so she can bring the empty dishes home to reduce clutter. The car trunk is always filled with several bags of dishes waiting to be cleaned. More frustrating than people's unkindness is the sheer number of cats helplessly roaming outside. Most of these cats were once somebody's companion but were abandoned or lost. Robin says that unless people spay/neuter all their cats and exercise responsible pet ownership, the feral/stray cat overpopulation problem will continue to grow.

"I don't want to be the person who turns away," Robin says. "I can't help every situation in the world but for me I just try to save every cat that I can. This is what I can do with my skills, with my time, with my energy and I can make a difference. I see a difference in the lives of those cats." [10]

Helder D.

Every morning before work, Helder visits five cat colonies to feed and water the feral cats. (He's been doing so for five years.) This all began on a cold winter morning, with the discovery of one homeless kitten. On the previous evening, the kitten got soaked in the rain. The temperature dropped during the night, so the kitten sought shelter in the seat of an abandoned car. In the morning, Helder discovered the kitten frozen to the seat of the car. Since that day, rain or shine, Helder diligently attends what has grown to five colonies to provide food, water, shelter and love for generations of feral and abandoned cats. He also traps the colony cats so they can be either spayed or neutered to help slowly reduce the colony population. For abandoned cats that have been forgotten by everyone else, Helder and his five colonies are the final refuge.

"I am hoping that a lot of people out there have the heart to join the cause. Call the Toronto Humane Society, call Toronto Street Cats . . . go in there and change the cages and feed the cats, whatever the case may be. Give them a hand and the world would be a much better place to live in."[11]

Francesca C.

Working from her garage turned recovery center, Francesca specializes in socializing abandoned and feral kittens. Here's the problem: non-spayed abandoned cats quickly become pregnant. Suddenly you've got one abandoned cat and half a dozen feral kittens with no human contact. Although they won't last long on the streets, they can be difficult to put up for adoption with no human contact. So Francesca hosts the cats and kittens in her garage where, in addition to tending to their basic needs, she also provides much needed love and socialization. This helps greatly to ease the transition from life on the streets to the warmth of their forever homes. Francesca has rescued and cared for many feral cats in York region. She is also a member of York Region Change for Ferals, an initiative to raise awareness, funds, and support for feral cats and those who care for them.

"You have to be compassionate to these living creatures because we are not the only ones living on this planet. We share this planet. Therefore, we need to work all together and make the best of it."[12]

Make no bones about it: people who care for free-ranging cats are well intentioned. They give generously of their time (and often of their pocketbooks) to care for creatures that many of us have chosen to ignore or, at worst, have intentionally and callously cast aside. To outside-cat advocates, the rights of the individual animal—each kitten, momma, or tom—are paramount. They want the cats to live, outside if that is the only option, and achieve the best level of "happiness" possible. They acknowledge that the problem begins with humans, and hope to undo a portion of the harm that some of our species have inflicted upon these animals.

Unfortunately, in their advocacy for outside cats, colony care-takers (and the organizations that support them) fail to take fully into consideration the health of the overall ecosystem and the rights of wild animals. Cats are opportunistic predators by nature. If given a chance to kill a bird or other small animal, most cats will take it. That is just the way cats are made. Maybe this chance arises once a day or once a week, and maybe they are successful only one-quarter of the time. But the fatalities add up to literally billions of amphibians, reptiles, birds, and mammals a year, which is enough to impact the well-being of whole species. Many cat advocates will aggressively contest the damage that free-ranging felines inflict upon bird populations. They will likewise deny the diseases that free-ranging cats spread to other mammals and even humans. But their hearsay and denials pale in the light of evidence of cat impacts on islands and the emerging hard science on their impacts on mainlands.

CHAPTER FOUR

The Science of Decline

> Few problems are less recognized, but more im-
> portant than, the accelerating disappearance of the
> earth's biological resources. In pushing other spe-
> cies to extinction, humanity is busy sawing off the
> limb on which it is perched.
>
> —Paul R. Ehrlich

Though he lacked the quantitative tools available to today's scientists, Edward Howe Forbush recognized the threat that free-
ranging cats posed to birdlife. Forbush was born in Massachusetts
in 1858 and grew up in Quincy, West Roxbury, and Worcester. He
was a passionate all-around naturalist and accomplished ornithologist from a very young age. Eastern Massachusetts was still heavily wooded in the mid-nineteenth century, and as a young child
Forbush immersed himself in the outdoors. At the age of fourteen
he taught himself taxidermy and by sixteen had been appointed
the curator of the Worcester Natural History Society's ornithology
collection. He was founder of the Massachusetts Audubon Society
and first president of the Northeastern Bird Banding Association
(later known as the Association of Field Ornithologists). He eventually became the Massachusetts state ornithologist. Forbush is
perhaps best known for *The Birds of Massachusetts*. It took him
four years to complete the three volumes; they were finally finished

in 1929, the year of his death. To this day, *The Birds of Massachusetts* is considered an extremely valuable reference on the birds of New England.

Forbush was a keen observer, and one of his responsibilities as an ornithologist was to document threats to birds. In 1916 he wrote a 112-page monograph entitled *The Domestic Cat: Bird Killer, Mouser and Destroyer of Wildlife; Means of Utilizing and Controlling It.* In his prefatory notes, Forbush provides some insight into his motivations for penning this monograph.

> Questions regarding the value or inutility of the domestic cat, and problems connected with limiting its more or less unwelcome outdoor activities, are causing much dissension. The discussion has reached an acute stage. Medical men, game protectors and bird lovers call on legislators to enact restrictive laws. Then ardent cat lovers rouse themselves for combat. In the excitement of partisanship many loose and ill-considered statements are made. Some recently published assertions for and against the cat, freely bandied about, have absolutely no foundation in fact. The author of this bulletin has been misquoted so much by partisans on both sides of the controversy that in writing a series of papers on the natural enemies of birds it has seemed best, in justice to the cat and its friends and foes, as well as to himself, to gather and publish obtainable facts regarding the economic position of the creature and the means for its control.[1]

At the time, most cities and towns in the Northeast were overrun with cats. To demonstrate the problem, Forbush collected statistics on the number of cats humanely destroyed by the Animal Rescue League of Boston and the Society for the Prevention of Cruelty to Animals based in New York. Over a ten-year period in Boston, 210,000 cats were destroyed, and in one notable day 269 cats and kittens were euthanized. In New York, the annual average of cats killed over ten years had been 16,400. But in 1911, the New York SPCA had killed over 300,000 cats. In both locations, the cats euthanized were largely animals taken from homes, rather than strays from the streets, so these numbers did not necessarily reflect the number of cats in the environment. Forbush saw an enormous problem unfolding. In another passage, he states the following:

> The widespread dissemination of cats in the woods and in the open or farming country, and the destruction of birds by them, is a much more important matter than most people suspect, and is not to be lightly put aside, as it has an important bearing on the welfare of the human race.[2]

Forbush likely knew little about the previous five mass extinctions on earth and certainly did not understand that he was in the midst of a sixth. But his observation is prescient.

What Forbush understood was that cats were spreading across New England like a veritable plague, and they were preying on enormous numbers of small birds and mammals. Forbush would also write about the diseases cats spread, especially rabies (see chapter 5). But the majority of his monograph focused on the direct impacts cats have on the environment, particularly their predation on birds. He conducted a survey all around New England and collected a series of anecdotal observations, and then used this information to estimate larger-scale impacts in a simple model. Some of the comments from people he queried include:

> I am skeptical when anyone says "my cat never catches birds," I have seen an active mother cat in one season devour the contents of almost every robin's nest in an orchard, even when tar, chicken wire and other preventatives were placed on the trunks of the trees.

> Mr Graham Forgie asserts that his cat kills about three birds per day.

> A friend had a cat which she was very proud of because it was such a good hunter and that in October it had killed and brought in twelve birds in two days. Nearly all of these birds were myrtle warblers.

Mr. Charles Crawford Gorst of Boston says that a friend told him that his cat had 14 birds laid out for its young for breakfast.[3]

Forbush liked to include quotes and anecdotes in his work—they go on and on, and each tells the same story. Forbush wanted to come up with an estimate of the numbers of birds killed per year by cats in the state of Massachusetts. He synthesized the information from all correspondents and estimated that each cat kills ten birds a year. He also estimated that on average there were two cats per farm. Based on this he estimated that in 1916, cats killed about 700,000 birds in the state of Massachusetts. Forbush considered this an underestimate, although some detractors found the number excessive. A colleague of Forbush, Dr. George Field, came up with an independent calculation. Field estimated that there was at least one stray cat for every 100 acres in Massachusetts, and that on average each cat kills one bird every ten days. This formula yielded an annual bird mortality estimate of 2 million for the state. Scientists from New York and Illinois did their own math using their best available data and came up with estimates of annual bird mortality due to cats at 3.5 million and 2,508,530, respectively. Forbush concludes his 1916 monograph: "The cat, an introduced animal, is not needed here outside of buildings. It has disturbed the biological balance and has become a destructive force among native birds and mammals."[4]

🐾

The year Forbush published his monograph on cats was a particularly transformational time for bird conservation. It was less than a decade after the back-to-back terms (1901–9) of Theodore Roosevelt. A powerful U.S. president, Roosevelt had a passion and dedication for understanding and conserving species along with their habitats. Much like Forbush, he was a lifelong naturalist with an insatiable curiosity for the outdoors. Roosevelt was quite knowledgeable about large mammals and was an advanced amateur ornithologist, keenly aware of issues surrounding bird conservation.

He kept a list of birds spotted on and around the White House property and collected and prepared many bird and mammal specimens (281 birds and 361 mammals, which now reside in the Smithsonian's National Museum of Natural History). His father, Theodore "Thee" Roosevelt Sr., whom he adored, was an ardent philanthropist and one of the founders of the American Museum of Natural History in New York. During his presidency, Roosevelt used his "bully pulpit" to protect close to 230 million acres of land by establishing 150 national forests, the first fifty-one federal bird reservations, five national parks, and the first four national game preserves.

Roosevelt knew of the extinctions of species such as the Great Auk and the Labrador Duck in the mid-1800s, and he had seen firsthand the demise of the American Bison, the Passenger Pigeon, and the Carolina Parakeet (and the eventual extinctions of the latter two in 1914 and 1918, respectively). Roosevelt's close friend Frank Chapman, considered a dean of American ornithology, made Roosevelt keenly aware of the massacres of waterbirds (herons, egrets, ibises) occurring in Florida and surrounding southern states, mostly for the purpose of gathering feathers to adorn women's hats. Populations of many species had declined to dangerously low numbers. Historian Douglas Brinkley, in his Roosevelt biography, *Wilderness Warrior: Theodore Roosevelt and the Crusade for America*, tried to capture his subject's feelings on the issue: "Certain bird species—herons, terns and ibises, for example—mesmerized Roosevelt. As president he insisted that killing one of these Florida exotics was a federal crime."[5]

The Roosevelt presidency put wildlife conservation on the map in the United States at a make-or-break moment in our conservation history. Roosevelt's close friendships with the likes of George Bird Grinnell, John Muir, and Gifford Pinchot (his eventual choice for the first chief of the U.S. Forest Service, 1905–10) profoundly influenced his outlook. In 1909, at the end of Roosevelt's second term, the North American Conservation Conference was held in Washington, and, at Roosevelt's urging, representatives from Canada, Newfoundland, and Mexico were present; Roosevelt likely understood the implications of the decline of the migratory bird species the nations all shared. By the end of the conference, a permanent conservation commission that contained members from each country had been established. This would eventually lead to a formal agreement between the United States and the United Kingdom (acting on behalf of Canada) to protect migratory birds. Signed on August 16, 1916, it was called the Migratory Bird Treaty. Two years later, in 1918, the U.S. Congress implemented the Migratory Bird Treaty Act (MBTA) to carry out the treaty's imperatives.

The MBTA provides that it is unlawful to "pursue, hunt, take, capture, kill, attempt to take, capture or kill, possess, offer for sale, sell, offer to purchase, purchase, deliver for shipment, ship, cause to be shipped, deliver for transportation, transport, cause to be transported, carry, or cause to be carried by any means whatever, receive for shipment, transportation or carriage, or export, at any time, or in any manner, any migratory bird, included in the terms of this Convention for the protection of migratory birds or any part, nest, or egg of any such bird." The MBTA today protects 800 species of birds and is probably the most important piece of legislation in the history of bird protection. That said, although it has worked to save several species of waterbirds from overhunting for feathers and food, it has not yet worked as a legal instrument to protect birds from cats (we will revisit this in chapter 6). This likely would have infuriated Forbush, who could see that cats were hunting and impacting many smaller and less charismatic species of birds (and mammals) than the humans hunting herons and egrets and that these species also needed federal protection.

Today, many species of waterbirds are flourishing. Herons, egrets, waterfowl, and numerous other groups have clearly benefited from the MBTA, from several other key pieces of legislation (e.g., the North American Wetland Conservation Act), and from the work of influential groups representing the interests of waterfowl hunters, such as Ducks Unlimited. But while the MBTA has worked for some species, it does not appear to have worked for all—especially nongame species. A recent analysis of bird population trends by a large assemblage of government, academic, and nonprofit groups, published in *The State of the Birds 2014*, suggests that populations of species considered to be indicators of wetland health have increased by 37 percent since 1968. Populations of indicator bird species from grassland habitats, however, have declined by over 40 percent, and some individual species, including McCown's Longspur and Sprague's Pipit, have declined by over 75 percent since 1968. Indicator bird species from eastern forests have declined on average by upwards of 32 percent, and some individual species, such as the Cerulean Warbler and Eastern Whip-poor-will, have declined by over 75 percent. Even species once thought to be common, such as the Rusty Blackbird, Common Nighthawk, and Chimney Swift, are disappearing before our very eyes. And shorebirds, seabirds, and all the endemic bird species of Hawaii are declining precipitously. Overall, in excess of one-third of the bird species (233 species) in North America have declined significantly since around 1970.

After forty-five years of bird population decline, it seems obvious that our currently available legal instruments are failing. More distressing is the fact that these declines are happening all over the globe, in similar patterns—wherever monitoring schemes have been in place, they are recording fewer and fewer birds year after year. In the United Kingdom, for example, woodland and farmland birds and seabirds, both migratory and resident, are all exhibiting declines similar to those reported in the United States. Roosevelt, were he alive today, likely would declare such bird declines, at home and abroad, a crime against humanity.

❖

You do not have to be a Nobel laureate or a president (or, as in the case of Roosevelt, both) to realize that species decline in number prior to going extinct. The fact that we are seeing such sharp declines in so many bird species likely means that other animal and plant species that we are not even monitoring are declining as well. Population declines can turn into extinctions quickly; a species that takes many tens of thousands of years to evolve can be gone in fifty years or much less (as in the case of the Stephen's Island Wren). Declining populations eventually reach a minimum size below which mates have difficulty finding each other, and even if they do, cannot produce enough young to sufficiently sustain or increase the population. Furthermore, if they do successfully reproduce, remaining individuals are often inbred and thus have severely reduced genetic diversity. Harmful mutations in offspring occur at higher frequencies, and the species is less likely to survive future environmental pressures (e.g., habitat loss or a new disease) should they arise. The Greater Prairie Chicken, Whooping Crane, Hawaiian Crow, and Florida Panther are all species (or subspecies) that have gone through population bottlenecks (small population sizes) and are experiencing reduced reproductive success and displaying higher rates of abnormalities. Although these species persist today, it is unclear how much longer they will remain on the planet.

One of our greatest challenges in the twenty-first century will be reversing the declines and potential extinctions of so many species. Part of the issue for bird extinctions lies in the fact that it is impossible to know where and how every bird dies—we do not have the equivalent of an airplane black box affixed to every individual. Sometimes the problem is clear, and this certainly was the case for a few waterbird species in the early 1900s—individuals were being overharvested (shot and killed) in tremendous numbers. People could see that big and charismatic animals like the Great Egret were disappearing. Cause and effect were relatively obvious, and there was little need for more information to spur action. The problem was less clear with the mid-twentieth-century disappearance of Peregrine Falcons, Brown Pelicans, and Bald Eagles, but once populations had dropped to precipitous levels, action was taken. The pesticide DDT was determined to be the cause, steps were taken (albeit slowly) to curb its usage, and the populations bounced back. In both cases, the problem was a single factor that once eliminated allowed species to rebound. Nature is resilient given the chance; once a threat is removed, populations often recover.

We do know human activities are largely responsible for the declines of bird species. Habitat loss, climate change, pesticides, collisions with large structures—as well as predation by cats—all cause mortality and play a role in driving species declines to varying degrees. We know that the impacts of some of these threats (a.k.a. mortality factors) are often indirect and not immediate. For example, habitat destruction and climate changes might impact reproductive success in a later season. We know as well that these threats can interact, that a given species can be impacted by more than one threat, and that impacts of multiple threats can accumulate. Take, for example, a New Hampshire–breeding population of the Ovenbird (*Seiurus aurocapilla*), a small songbird from the warbler family. Its annual cycle extends from those breeding grounds in New Hampshire, where it resides from May to July, to wintering grounds in Cuba and the Dominican Republic, where it lives from October to April. Birds from that population can die at any time while on the breeding or wintering grounds, or during their roughly month-long migrations to and from these areas. It is difficult to identify with any precision the relative impact of a mortality factor, such as the free-ranging cat, to all birds that make these journeys over such large spatial areas and wide expanses of time—and most bird species (>75 percent) in North America migrate! We count birds and detect population declines while birds are on their breeding grounds, yet the mortality can occur at any time of the year. The causation mystery is complicated by the fact that when a small- to medium-size bird (say, a sparrow) dies in the wild, it often vanishes within hours. If it is killed by a predator, mortality is swift, and the prey is usually consumed rapidly. A pile of feathers is all that is left. Few people see the death of an animal in nature.

The fact that cats kill small animals like birds and mammals is by itself not groundbreaking news (fig. 4.1). The challenging issue, and the bar that has been set as to whether we tolerate outdoor cats on the landscape, has been to determine whether cats impact not just individuals of a given species but the broader populations within which individuals exist. Natural populations fluctuate in size from year to year and must maintain a certain number of individuals to remain stable over the long term. How do we assess the magnitude of cats' effect on naturally fluctuating wildlife populations? This requires understanding whether mortality from cat predation is "compensatory"—meaning these deaths substitute for deaths that would have occurred from other causes like disease or starvation—or "additive," meaning deaths that add to the tally of those caused by other factors. Some argue that we should not be concerned about the mortality imposed by cats because it is compensatory: the animals killed by cats are from a "doomed surplus" that would have died anyway. However, if increased mortality due to cats is additive, stable populations would start to decline, and already declining populations would spiral downward even faster.

Demonstrating whether a given mortality event is compensatory or additive over a large area, such as a country or even a state, is very challenging; it is simply impossible to track all mortality events and assign causation to each incident within a population. Tracking mortality within migratory bird populations poses an additional complication: In most cases we do not know where breeding birds from a particular area go to spend the winter, nor where wintering birds go to breed in the summer. When a northbound migratory bird like an Ovenbird rests and refuels within a patch of woods in Cape May, New Jersey, in the spring, only to get killed by a cat, we do not know the breeding population to which that bird was going or the winter population from which it came. This obscures our understanding of impacts on population processes of these populations. But there are ways to assess the degree to which cats may impact important population processes, including vital rates like survival and reproduction at local scales that can illuminate population declines at larger scales.

☙

There are an estimated 8.1 million domestic cats in the United Kingdom today, a particularly high density of cats for a land area the size of Alabama. Most of these cats are owned. The English are by and large a nature-loving people; more than 1 million people claim membership to the Royal Society for the Protection of Birds (RSPB), the British equivalent of the National Audubon Society. Yet the English are inclined to let their cats roam outdoors, and even the RSPB has an article posted on its website stating that free-ranging cats are not causing a problem and that their impacts represent compensatory mortality.

Peter Churcher and John Lawton knew the challenges of demonstrating whether a particular source of mortality is additive versus compensatory—so they set up a study explicitly focused on this problem in a small town—the Bedfordshire village of Felmersham, about sixty miles due north of London. Their study tracked seventy domestic cats and all prey items each cat brought back to their homes over the entire year of 1981. In total, cats brought back 1,090 prey items: 535 mammals, 297 birds, and 258 unidentified animals. House Sparrows (*Passer domesticus*)—a native resident species in England and Europe—made up 16 percent of all kills. Churcher and Lawton also estimated the population of House Sparrows over the same area where they were tracking cats. This allowed them to gauge the specific impact of cats on the population of House Sparrows in the area where the two overlapped. They estimated that cat kills made up a minimum of 30 percent, and possibly as much as 50 percent, of the annual mortality for this bird. These figures led them to conclude that cats were imposing significant predation on House Sparrows and implied that they likely were "additive" to the natural levels of mortality. House Sparrows declined by over 60 percent in England from 1994 to 2004, and although recent experimental evidence suggests that nestling food limitation was the major limiting factor, it is not too much of an intellectual stretch given Churcher and Lawton's findings to see that predation by cats is also a significant contributing factor.

Understanding the mortality impacts of cats does not always require putting the research into the framework of additive versus compensatory mortality. A study by Kevin Crooks and Michael Soule in 1999 provides another clear illustration of how domestic cats drive declines and local extinctions of songbird populations. Their study took place in the native coastal sage-scrub habitats of southern California, where land has been fragmented severely by varying degrees of development. This habitat fragmentation has resulted in the elimination of the Coyote (*Canis latrans*) in some habitat patches, while the species' populations have remained relatively intact in others. This patchiness allowed the researchers to test an idea known as the "mesopredator release hypothesis." The hypothesis predicts that the absence of a large predator (Coyotes) results in the release or population growth of mesopredators, such midsize predatory animals as Raccoons, opossums, and cats. Because midsize predators prey more commonly on songbirds, the hypothesis also predicts a decline and even local extinction of birds and other small prey items. As predicted, where Coyotes were absent, numbers of cats and Raccoons went up, and bird abundance and diversity declined. Similarly, where Coyotes were present and abundant, mesopredators were absent, and bird diversity and abundance was high. In other words, Coyote and cat abundances were better predictors of bird abundance and diversity than other habitat-related factors. Crooks and Soule also demonstrated that Coyotes eat cats when they are present in the habitat fragments. Cat remains showed up in 21 percent of all Coyote scat samples. A more recent independent confirmation of how much Coyotes like to eat cats comes from North Carolina–based zoologist Roland Kays and colleagues. Using camera traps operated by citizen scientists within thirty-two protected forested areas in six states along the East Coast, they found that when Coyotes were present, cats were absent; and when Coyotes were absent, cats were present.

Perhaps even more noteworthy in the Crooks and Soule study is their demonstration of how predation by cats, because they are subsidized predators, can be so much more significant compared to natural predation levels. They estimated that residents with houses bordering forest fragments each owned on average 1.7 cats. Seventy-seven percent of these owners let their cats outdoors, and 84 percent of these cats brought back kills. The researchers went on to calculate that for each moderately sized land fragment (50 acres), there were approximately thirty-four bird-killing cats. By comparison, natural predators (Raccoon, skunks, Coyote) in each fragment would typically be present in no more than one or two pairs. How can a fragment support so many cats—and even when its birds are eliminated, further lowering food availability for predators? It is simple—the cats are not killing for food. Owned cats (and colony cats) are subsidized predators—their persistence is not determined by the food they gather; they are already fed at home with a can of tuna-flavored Friskies Buffet. This subsidization allows free-ranging cats to persist at levels far above natural predator densities and the carrying capacity of a particular habitat. As a result, their impact can be far greater than that of any natural predator. There is no shortage of studies documenting the rates of predation at local and regional scales of cats on a diversity of bird and mammal species, including California Quail, California Thrasher, Gray Catbird, Northern Mockingbird, Black Redstart, Wood Mouse, and harvest mice.

Sometimes even the mere presence of a predator can impact prey species. Colin Bonnington and colleagues, of the Department of Animal and Plant Sciences at the University of Sheffield, England, for example, had a hunch that the presence of a cat near a bird nest could have indirect, sublethal effects on individuals and populations. They knew from previous research that predators influence prey populations simply by altering prey behavior—such effects are exceedingly common across many predator-prey systems. Such sublethal effects can reduce population size by changing the habitats an animal might use and also by reducing parental care at a nest—both of which impact reproductive success. Bonnington and colleagues tested the latter idea in a suburban area of Sheffield during the breeding season of 2010 and 2011. They found blackbird nests and placed a taxidermy mount of a cat, a squirrel, or a rabbit within about six feet of the nest. The mounts were left in place for only fifteen minutes; once they were removed the researchers measured the rates at which the adults attended to the nest, either to commence incubation or to feed nestlings. The results were clear. In response to the cat mount, adults made significantly more alarm calls and reduced their provisioning rates by up to one-third, compared to the adults reacting to the rabbit and squirrel mounts. In addition, because the adults reduced their nest attentiveness, significantly more nestlings were eaten by predators after exposure to the stuffed cat compared to exposure to the stuffed rabbit or squirrel. Changing the predation rates at nests can be a major contributing factor to whether a population grows or declines.

It is clear there are many ways, including both through direct mortality at local scales and indirect sublethal impacts, that cats can influence the behavior of prey and populations of birds and mammals. Understanding the impacts of cats at large scales—specifically, how they might be contributing to population declines of widespread species—is challenging, as we have discussed. But the local-scale impacts—which, after all, add up to large-scale impacts—send a clear message: cats impact bird and mammal populations. That being said, can we estimate the amount of mortality that cats impose at even larger scales, say across the United States?

Once again we need to turn to statistical modeling, and a first step is to physically count the number of dead birds brought home by a cat (if not already consumed). Such tallies can provide important insights into the relative severity of predation by cats. At least fifty-five independent, peer-reviewed studies have quantified the rates at which species such as amphibians, reptiles, birds, and mammals are killed by owned and unowned free-ranging cats. After excluding studies with unusually high mortality estimates, small sample sizes of cats (< 10), those with experimental manipulations of any kind, or those involving questionnaires asking people to recall predation events, Scott Loss, Tom Will, and Peter Marra found seventeen studies of owned cats allowed to roam outside in which the animals brought in a minimum of 1.14, and a maximum of 33.18, birds per cat per year. Forbush was not that far off when he estimated that owned cats in Massachusetts brought in about ten birds per year each. Nineteen published studies of the number of prey killed by unowned cats found that these animals returned between 30.0 and 47.6 birds per cat per year. Rates of mammal kills per cat per year were found to be much higher, with estimates between 8.7 and 21.8 mammals for owned cats and between 177.3 and 299.5 per unowned cat per year.

These numbers are almost certainly underestimates, as cats do not bring back everything they kill. Two studies confirm that they hold back some of their quarry. Roland Kays and Amielle DeWan, in 2004, put radio collars on eleven owned cats around the Albany, New York, area and asked their owners to keep all the prey their cats brought home so positive identifications could be made (fig. 4.2). They also did intensive observations of the cats in the field to quantify actual kills. Cats brought home about 1.7 prey items per month but actually killed 5.5 prey animals each month—more than a threefold difference. About ten years later another study, this one led by graduate student Kerrie Anne Loyd at the University of Georgia, looked at the same problem but with a new twist—she employed the use of "Kitty Cams." These are small video cameras suspended around the neck and just below the chin of a cat for up to ten days. Cat owners removed the cameras at the end of each day to download data and recharge the Kitty Cam battery. Loyd collected data from fifty-five cats over a one-year period. The video footage showed that 44 percent of the cats (roughly twenty-two animals) hunted wildlife, and that these cats brought home less than 25 percent of what they actually killed. The implication of both studies is that prey-return data underestimate actual kill rates by owned cats by a significant amount.

If you can estimate the number of animals killed by cats over some time frame, and you know the number of cats that hunt (as Forbush did in 1916 in Massachusetts and as Temple did in 1986 in Wisconsin), you can estimate the total number of animals killed by cats at the scale of interest. In 1991 a conservationist and ornithologist named Rich Stallcup attempted such a tally. Stallcup was a legendary birder and all-around naturalist from Oakland, California. He cofounded the renowned Point Reyes Bird Observatory in 1965 (PRBO, now Point Blue Conservation Science) and was famous for his uncanny ability to bring nature to humans. Over his career, Stallcup wrote several books and seventy-five articles for his "Focus" column in the *Quarterly Journal of the Point Reyes Bird Observatory*. One such column, published in 1991 on the heels of the Temple estimate for Wisconsin, generated great attention. Entitled "Cats: A Heavy Toll on Songbirds. A Reversible Catastrophe," it was one of the first (albeit completely back of the envelope) estimates for the bird mortality caused by owned cats across the contiguous United States. Stallcup described how songbirds on every continent are in steep decline and for many reasons—global warming, habitat loss, but also *cats*. His point was clear: the cat take across the United States was massive, and compared to so many of the other threats facing birds, the cat problem was reversible.

Stallcup did some simple calculations to develop his model. First he needed to estimate the number of owned cats that were allowed outside. He used an estimate from the *San Francisco Chronicle* (March 3, 1990) that put the number of domestic house cats allowed outside at 55 million. Stallcup viewed this as a conservative estimate but reduced this number by 20 percent, assuming that at least that many cats were not let outside or were too old or slow to catch wildlife. This gave him a final estimate of 44 million owned cats that were allowed outside and able to hunt. He then estimated the kill rate of the 44 million cats. He wanted to be "very conservative," so he estimated that only one out of ten cats kills one bird a day resulting in 4.4 million birds per day, or easily over 1 billion birds killed by cats per year across the United States. One could argue that Stallcup's numbers were underestimates for multiple reasons—including the fact that he did not include feral or free-ranging, unowned cats in his population estimates. In his words:

> Add to this the plague of feral cats. How many? No one knows, but they occur everywhere in temperate North America (except deserts and high mountains), and in some places are abundant. . . . Along the California coast it is common to see 10 to 15 during a day's outing (and these are nocturnal animals). Certainly, there are many million, country-wide. What do they eat? Wildlife. Nothing but wildlife.[6]

Despite Stallcup's rough estimate, from the early 1990s, the birding establishment endorsed a much lower number. In ornithological textbooks (such as the most recent edition of *Ornithology* by Frank Gill) and field guides (including those published by famed birder and artist David Sibley) the number listed was hundreds of millions (up to 500 million) of birds killed by cats every year. It was as if even ornithologists refused to believe the number could be so high, even though the math was straightforward and simple. Interestingly, even at just 500 million birds killed by cats each year, cat-caused mortality was considered to be the second-highest cause of bird mortality, after window strikes—collisions with glass windows on buildings and houses—which were thought to kill 1 billion birds per year. Because it had not really been done before, a U.S.–wide estimate, developed from the most solid peer-reviewed science available, was needed to better understand the magnitude of cat impacts. By 2013, hundreds of papers had been published to support the development of this more exacting analysis. Such an estimate also needed to incorporate uncertainty (a minimum and a maximum kill rate) for the number of birds killed by free-ranging cats every year; remember, there is uncertainty involved even in estimating something as simple as how much it costs to drive the length of the state of California.

A simple modeling exercise was precisely what Scott Loss, Tom Will, and Peter Marra did with their groundbreaking study in 2013. The estimate they developed for cat-caused animal mortality was part of a larger effort to develop better estimates for each of several direct but unintended human-caused mortality factors for birds in the contiguous United States. These included collisions with buildings (primarily windows), communication towers, wind turbines, and vehicles; electrocutions at power lines; and predation by cats. Existing figures were in need of checking and updating; and in the case of cats, a first systematic estimate was needed. For mortality due to free-ranging cats, Loss et al. first performed an exhaustive review of the existing scientific literature for studies that included cat-caused mortality on amphibians, reptiles, birds, and mammals to come up with the best numbers to plug into their model. They looked at hundreds of studies but included in their review only those from temperate mainland or large island areas (New Zealand and the United Kingdom), with a sample size of at least ten cats and at least one month of sampling. To reduce bias, studies were excluded from the final analysis if they, for example, reported unusually high estimates of mortality, or if the cats wore bells or bibs that may have reduced predation rates.

The final model for estimating total annual mortality was a little more complicated than previous versions (e.g., Stan Temple's) but still straightforward. The researchers first estimated the annual mortality from owned cats with this formula:

$$\text{Annual mortality from owned cats (mp)} =$$
$$npc \times pod \times pph \times ppr \times cor$$

- *npc* is the number of owned cats in the contiguous United States
- *pod* is the proportion of owned cats allowed to have outdoor access
- *pph* is the proportion of outdoor owned cats that hunt wildlife
- *ppr* is the annual predation rate by outdoor-hunting owned cats
- *cor* is a correction factor to account for the fact that owned cats do not return all prey to owners (remember that cats give their owners only a portion of what they kill)

Then, they estimated the annual mortality from unowned cats:

$$\text{Annual mortality from unowned cats (mf)} =$$
$$nfc \times pfh \times fpr$$

- *nfc* is the number of unowned cats (or free-ranging cats) in the contiguous United States
- *pfh* is the proportion of unowned cats that hunt wildlife
- *fpr* is the annual predation rate by hunting unowned cats

The total annual mortality is the combined annual mortality estimate from owned and unowned cats:

$$\text{Total annual mortality from all cats} = mp + mf$$

Let's consider how Loss et al. came up with the numbers for each of the figures in the model in a little more detail. First, how many cats are there in the United States?

Fortunately, at least two estimates for the number of owned cats in the United States had been developed since Stallcup gleaned his information from the *Chronicle*. Two independent nationwide pet-owner surveys estimated the number as 86.4 million and 81.7

million cats, respectively. The mean estimate of 84 million owned cats is almost twice as high as Stallcup (or at least the *Chronicle*) had reported just twenty years earlier. Next: how many of these cats were allowed outside, and how many then hunted? Based on eight different studies, between 40 percent and 70 percent of owned cats were allowed outside; three additional studies suggested that between 50 percent and 80 percent of these animals actually hunted. Seventeen peer-reviewed studies were used to estimate the bird return rates per owned cat per year. An additional twenty-six studies were used to estimate mortality rates on amphibians, reptiles, and mammals. A correction factor (between 1.2 and 3.3) was then included in the model to account for the fact that cats do not always bring their kills home. Calculating the model estimates for unowned cats was a bit more complicated.

Actual estimates of the number of outdoor unowned cats just do not exist, for several reasons. First, cats are not easy to detect and count. They are quiet, stealthy, and intentionally avoid notice—all evolved behaviors consistent across all cat species. The other problem is that people who maintain colonies of cats do not report their whereabouts and do not keep records of their numbers—despite calling them "managed colonies." Rough estimates do exist and include between 20 and 120 million unowned outdoor cats, with 60 to 100 million cats the most frequently cited range. Because of this uncertainty, Loss et al. used a minimum and maximum of 30 million and 80 million unowned cats—a very conservative number. Studies of unowned cats typically report that 100 percent are hunters, so this parameter was set between 80 percent and 100 percent. Finally, from a total of forty-five peer-reviewed studies conducted in temperate regions, Loss et al. estimated that each individual unowned cat annually kills 1.9 to 4.7 amphibians, 4.2 to 12.4 reptiles, 30.0 to 47.6 birds, and 177.3 to 299.5 mammals per year. Given this component data, coming up with the final estimates for the magnitude of the impacts of free-ranging cats was now as simple as pushing a button.

Pushing a button, or buttons, is exactly what happened. No one had ever tried to rigorously quantify, through a synthesis of the best available data, the number of animals killed by cats across the

United States. The numbers were substantially higher than anyone expected, especially given that previous estimates for birds had been in the hundreds of millions (except for Stallcup's), and there had been no estimates for other animals. The final mortality numbers showed that cats killed between 1.3 and 4 billion (median 2.4 billion) birds per year, with unowned cats causing the majority of the mortality (69 percent). Many of these birds were likely juveniles, but no details of species or age and sex were available. The final estimates for mammal mortality were also alarming; 6.3 to 22.3 billion (median 12.3 billion) mammals were killed every year by outdoor cats. Annual mortality for amphibians and reptiles was in the hundreds of millions—95 to 299 million amphibians and 258 to 822 million reptiles. Even the minimum estimates, which are highly conservative, were cause for double takes.

As shockingly high as the numbers are, the analysis is sound; the article announcing the results was reviewed by some of America's most accomplished scientists. It is interesting to note that the study dovetailed closely with a similar analysis of direct anthropogenic mortality of birds conducted in Canada, also in 2013. Across Canada, where there are fewer cats than in the United States, the animals were implicated in the deaths of 204 million birds per year (a median estimate). As in the United States, this makes cats the most significant source of direct anthropogenic bird mortality.

Do these estimates allow us to say with certainty that cat mortality at continental scales is additive or compensatory for bird populations? No, they do not. For the reasons described earlier, these estimates cannot definitively answer that question because information is, and will continue to be, incomplete at these scales—and also because we do not have reliable estimates for the population sizes of most bird species in the United States. For small mammals, reptiles, and amphibians, estimates of population size do not even exist. What the estimates do provide is insight into the magnitude of the mortality. And when combined with the many local studies that illustrate impacts on population processes (as we have described above) at smaller scales, they raise a serious concern about the ecological impacts of free-ranging cats.

❧

The Loss, Will, and Marra paper appeared in the international science journal *Nature Communications* on January 29, 2013. That same day, a Tuesday, *The New York Times* picked up the story. Running under the headline "That Cuddly Kitty Is Deadlier Than You Think," the piece, by science reporter Natalie Angier (which included a photo of a domestic cat clutching a rabbit in its jaws) ignited a firestorm. The science and the various interpretations of the science emerged from the obscurity of the cat and ornithology communities and landed squarely in pop culture. Angier's piece was the most e-mailed and most commented-upon piece that week on the *Times*'s website. It was more popular than stories on the war in Afghanistan, the world economy, and human poverty. Within twenty-four hours, more than 300 other international media outlets (including NPR, *USA Today*, the BBC, and the CBC) picked up the story, and approximately 600 million unique viewers on websites read the report that estimated that kill rates by cats are three to four times higher than mortality figures previously bandied about. The Loss et al. paper positioned the domestic cat as one of the single greatest human-linked and direct threats to wildlife in the United States, and emphasized that more birds and mammals die at the mouths of cats than from wind turbines, automobile strikes, pesticides and poisons, collisions with skyscrapers and windows, and other so-called direct anthropogenic causes combined.

The *Nature Communications* paper struck a chord. Cat lovers and bird lovers—already at odds—finally had a public spotlight for debate. The battle lines were bright and clear. Some were in favor of leash laws for cats, euthanasia of strays, and the elimination of cat colonies and trap-neuter-return (TNR) programs (in which unowned cats are caught, vaccinated, spayed or neutered, and released into the wild again). Others were in favor of leaving any and all cats free to be, kill, and cuddle. A brief survey of the 1,691 comments the *New York Times* received on its website over three days illustrates the polarizing topic:

Simply keep cats inside. Indoor cats are healthier, less prone to disease, fleas and mites, and are much more sociable. Particularly in the suburbs and rural areas, mice, voles, moles, snakes, amphibians and birds must be protected from cat predation. These native creatures are valuable pollinators and seed dispersers for native plants as well as food for native predators . . .

I have maintained a feral colony for 15 years in Austin TX. There were close to 30-40 cats in the colony when I started and moved in the home near their territory. I fed them, made sure they had access to water, trapped, neutered and released every member, fixed every kitten and found homes for them if the queens were not spayed in time. Some cats came, some cats went. After 15 years, there are now only 3 left (sad face). The oldest is 9 years old and there were 2 recent additions. The old one is friendly now to me and will stay on as an outdoor kitty. The 2 new ones are friendly and will be fixed, vaccinated and rehomed soon. While I don't see many dead birds, I do find dead rodents and snakes (mostly coral and rattler). Trap, neuter, release does work over time and WHAT would you rather have near your home? DISEASE CARRYING MICE AND RATS, POISONOUS SNAKES, or a few feral cats that leave you alone?

I'm surprised by all of these cat friendly comments. I love cats and have two who stay entirely indoors, and as far as I can tell don't really want to go outside. There are many species of native birds that are endangered, partially due to predation by cats, a non-native species. Cats should not be allowed outside, period. It is irresponsible, and will eventually result in the extinction of more native species. In my opinion, feral cats should be humanely killed. I love cats, but I love birds too, and can't stand to see my neighbors' cats roaming my neighborhood, looking for their next kills.

Here in Florida we encourage our cat to roam and capture rats and mice. We encourage our cat (Rusty) to ignore cat haters who keep their cats in cages or trapped in a house with nothing but a sand box and a scratching post. I read your piece to him and he appears not the least put out and commented that cats that can't hunt are like people who prefer canned food over fresh.[7]

A majority of ecologists, ornithologists, and millions of bird aficionados see outdoor cats, whether owned or unowned, as killing machines. Many biologists are convinced that predation by this invasive species is indeed contributing to the catastrophic downward spiral of many bird and mammal populations. The tens of thousands of well-meaning people who nurture unowned cats, and the millions of domestic-cat owners who let their cats outdoors, all value these animals as sentient beings. They view them as part of the landscape, as much an element of the natural order as trees and clouds. Some in the cat advocacy world say, "We are a nation of animal lovers. We are not a nation of cat people or bird people." Yet there is a conflict between cat advocates and bird advocates—a war, quite literally to the death in the animals' case, whether or not the cat lovers or bird lovers will admit it.

Stepping back, what makes the impacts of cats especially troubling is that today every species that goes extinct or is declining in population size is doing so because of human activities. To a large degree, we control how quickly or slowly a species goes extinct. The problem is that the rate at which species are disappearing from the planet is far more rapid than what used to be considered a natural or background rate for species extinctions to occur. The background rate for species extinction is estimated by analyzing fossil records for millions of years prior to human existence (estimates are calculated as extinctions per million species-years, or E/MSY). Basically, the natural rate is estimated at roughly two extinctions per 1 million species-years. Put another way—there are two extinctions per 100 years per 10,000 species. This has led many scientists to conclude that this period of human existence, otherwise known as the Age of the Anthropocene, is the sixth mass extinction in the history of life on earth.

Since life on earth began approximately 3.4 billion years ago, more than 5 billion species have emerged, evolving into a vast diversity of taxa from viruses to dinosaurs to cats. Most of these species, upwards of 99 percent of all that once lived on earth, have now gone extinct. The vast majority of these species extinctions occurred during five distinct and significant prehistoric events, recorded in the fossil record, which dates back 450 million years. The first of these major events began 447 million years ago, in the period of time known as the Ordovician, when all known life is believed to have occurred in the oceans. Then the climate began to change and continued to change over a 4-million-year period. It got cold, extremely cold, especially in the Southern Hemisphere, where rich coral reefs and associated species such as nautiloids, trilobites, and brachiopods became locked in ice, which caused the extinctions of these and eventually most other marine species. Large sheets of ice eventually blanketed the entire southern continent of Gondwana. (At this point in the earth's history, the land was separated into two large supercontinents—Gondwana in the south and Laurasia in the north.) As water became locked into ice, sea levels farther north declined, causing water chemistry to change. This further amplified the number of species extinctions in other parts of the globe. Eventually an estimated 75 percent of all species on earth went extinct in this period due to climate change; it would be the second-greatest mass extinction in earth's known history.

Four more mass species extinctions subsequently occurred: the Ordovician–Silurian extinction, the Permian–Triassic extinction (the largest), the Late Devonian extinction, and, the most recent, the Cretaceous–Paleogene extinction. The lattermost occurred just 66 million years ago. Each of these events had various causes, ranging from a gigantic asteroid striking the earth, a sudden release of methane gas from the ocean floor, climate change, or a combination of the above. These are all theories, of course, but most are based on strong lines of evidence. The mass extinctions themselves are not theories—they are fact. Another fairly certain fact is that humans, had they existed, likely would have also gone extinct during any of these catastrophic events.

In the previous five mass extinctions, events like comets crashing into the earth and eruptions of methane gas from the ocean floor were out of the control of any living thing. Dinosaurs were minding their own business when, *BAM*—here's an asteroid! In this, the sixth mass extinction, human overpopulation and subsequent habitat destruction and climate change are clearly the primary drivers, but there are others—and the effects are cumulative and interacting. Human overpopulation, for example, results in overharvesting (fishing, hunting), various forms of pollution, and the spread and maintenance of invasive species. Cats are one invasive species in particular that, according to Felix Medina and his colleagues, has been implicated in at least 14 percent of global reptile, bird, and mammal extinctions on islands (discussed in chapter 2).

Cats are clearly having an impact and in that sense contributing to the sixth mass extinction. Are they the primary driver? No. But we cannot afford to focus solely on the main driver. We must tackle all the component parts and certainly the ones we have the power to control. If we were talking about the well-being of people instead of birds and were to focus only on solving the primary cause of human mortality, we would ignore all cancers, AIDS, drunk driving, and a host of other health and social maladies and instead focus all of our efforts on heart disease alone. This would hardly be acceptable.

We have known for many years that cats can cause extinctions and have significant impacts on birds and other small animals on islands and mainland areas. Countless studies have been conducted worldwide to document the impacts of cats at different scales. Collectively, the science is overwhelmingly conclusive that cats kill massive numbers of birds and other small animals and that these deaths impact population processes. Moreover, cats are known to have contributed to the declines of many island species and subspecies that have not gone completely extinct yet, from the Hawaiian Crow to the Socorro Mockingbird to the Lower Keys Rabbit (the list goes on and on). Many remain skeptical that cats are having a significant impact on wildlife, especially on a continental level. They argue that we are not seeing actual species extinctions or even a demonstrable population decline that can be tied to cats. Although information is not complete for tying the population decline of a given species to cats at large scales, information describing cats' impacts on birds and other animals is clear and, when used in models, points to a need for action. The need for action is made more convincing by the critical point that extinctions are just one metric of global environmental health. When we lose a species to extinction, or an entire species declines, or local populations are wiped out, we are losing the important ecological functioning and critical ecosystem services that each of those populations provides. Collectively these extinctions and declines are all events that are contributing to the sixth mass extinction.

And if it is not bad enough that outside cats are accelerating the extinctions of many species of wild animals, they are also, evidence strongly suggests, as we will see in the next chapter, sickening—and in many cases killing—humans.

CHAPTER FIVE

The Zombie Maker:
Cats as Agents of Disease

> We stopped looking for monsters under our bed
> when we realized that they were inside us.
> —Charles Darwin

As beautiful and deadly as predator-prey dynamics are, host-pathogen interactions take the Oscar for brilliant orchestration and level of devastation. They are literally the inspiration for Hollywood movies. In fact, Hollywood has a long history of pulling from this biological genre, because it feeds the imaginations of millions of people and pays big dividends at the box office. Think *Invasion of the Body Snatchers*, *The Thing*, *Contagion*, or any of the dozen or so zombie movies that have graced the big screen. The premise is simple—a pathogenic organism enters a human body and either kills its host outright or changes its behavior to do equally horrid things. Although some of these may be over the top (e.g., with body-invading organisms originating from outer space), equally disturbing creatures exist here on earth, right before our very eyes—or perhaps in them (more on this later).

While much of the Hollywood spin is fictitious, there is a kernel of truth to these movies that keeps us coming back to theaters and obsessively squirting gobs of antimicrobial soap into our palms. The threat of a mysterious organism invading our bodies—perhaps jumping from a rat, bat, bird, or cat rather than from outer space—is not just compelling, it is real. A zoonotic disease, or zoonosis, is a sickness that emerges when a pathogenic organism, such as a virus, bacterium, protozoan, or fungus, invades a human body from another animal species. Over the ages, zoonoses have been responsible for the deaths of hundreds of millions if not billions of people.

Imagine waking up with a collection of large, apple-size bulbous swellings in your groin or armpit that secrete blood and pus. Eventually, black spots start spreading all over your body, unbearable fever sets in, you start vomiting blood, and then—usually within two to seven days of the onset of symptoms—you die. The zoonotic disease responsible for these horrific symptoms, black plague, originated in China in the early to mid-1300s and then spread to Europe and the Middle East. It eventually killed about one-third of the European population—somewhere between 75 million and 200 million people. At least one more major plague pandemic would occur in the mid-nineteenth century, again starting in China, then making landfall in San Francisco and eventually going global. Today regular outbreaks of plague still occur in Africa and China, and every year in the United States there are ten to twenty cases.

The plague pathogen is a small rod-shaped bacterium called *Yersinia pestis* that is transmitted by a "vector," the organism that transfers a pathogen into a host. Plague is primarily transmitted by a flea vector, whose hosts include more than 200 small rodent species such as the Great Gerbil, the Black Rat, ground squirrels, prairie dogs, chipmunks, marmots, and several other mammal hosts. Three clinical forms of plague typically emerge in humans: bubonic (the most common), pneumonic (pulmonary), and, the rarest form, septicemic, which infects the blood. In some host species, like cats, the bacteria are more likely to settle in the lungs. When this happens, cats spread the more deadly pneumonic form of the disease. People typically, but not always, contract the plague when infected fleas jump ship from a primary host and bite humans (who are known as secondary hosts, because the pathogen only invades them briefly while transitioning to its next life stage). But cat-to-human and human-to-human transmission through aerosol droplets can also lead to the deadly pneumonic plague. Finally, plague can also be contracted by eating plague-infected meat (think guinea pigs in Peru and Ecuador). Whatever the mode of transmission, after a two- to seven-day incubation period, symptoms set in and, if left untreated, death is certain and swift. If treated, approximately 50 percent of bubonic plague patients survive, but very few pneumonic or septicemic plague patients do.

It is doubtful that "John Doe," a thirty-one-year-old man visiting Chaffee County, Colorado, had plague on his mind as he made his way into the crawl space of a house he was visiting on August 19, 1992, to catch a neighbor's cat. Chaffee County sits almost smackdab in the center of Colorado: it is a rural and mountainous area with a low human-population density. The cat died minutes after being brought outside. It did not occur to John Doe at the time that he should be concerned for his welfare. When interviewed days later, the owners of the cat said they had noticed abscesses, lesions, and blood-tinged sputum on the animal—all symptoms of plague infection in cats. John Doe returned to his Pima County, Arizona, home and three days later, on August 22, began feeling abdominal cramps. The following day he had a fever—his temperature climbed to 103°F—along with vomiting and diarrhea. His condition worsened, and he was admitted to a hospital on August 25. He died within twenty-four hours. Postmortem tests confirmed the causative agent of death—*Yersinia pestis*—the same pathogen responsible for the black plague and millions of deaths in medieval Europe. Searches around the Colorado house for infected rodents and fleas turned up a dead Colorado Chipmunk that tested positive for *Y. pestis*—perhaps the cat's prey from the week before. John Doe apparently had had enough face-to-face exposure with the cat during the brief extraction from the basement for respiratory droplets to make their way into his body. The bacteria they carried killed him within a week.

Cases of plague transmitted from cats to humans are rare in the United States. From 1977 to 1998 there were twenty-three cases of cat-associated human plague in the country. At least one case occurred per year in a swath of eight western states, five of which were fatal (including John Doe's), either because they were diagnosed too late or simply misdiagnosed. Cats transmitted plague to their owners, caregivers, and veterinarians via bites, scratches, aerosol droplets, and through the simple act of curling up on a lap and purring near the owner's face. While most vector-borne diseases are seasonal in occurrence, plague in cats is not. These cases occurred in every month of the year but January and February and often were not associated with outbreaks of plague in nearby rodent populations. The mammal species that cats prey upon are ideal reservoirs for plague because they may remain asymptomatic throughout the year and therefore stay healthy enough to be able to transmit the pathogen if attacked or killed by a predator, which contracts the disease either through eating the infected prey or from the prey's fleas. Plague is endemic in seventeen western U.S. states, and if you live in certain areas there (typically more rural regions with rodent populations) and you own, handle, care for, or treat an outdoor cat, you need to be vigilant for plague. In fact, wherever and whenever cats are allowed outside their owners need to be on guard for numerous disease-causing agents—many of which may not simply sicken or kill a cat but will also sicken or kill other species of wildlife—and humans.

"Cat scratch fever" can mean different things to different people depending on their frame of reference. The rock musician Ted Nugent made the phrase famous in 1977 when he released a song that metaphorically equated the disease to a man's feverish desire for a female. More commonly (and perhaps more appropriately), cat scratch fever refers to an infection from a *Bartonella* bacterium that develops when an infected cat scratches or bites the human skin. In cats themselves it is usually not a serious problem, and 40 percent or so of the cats that carry it are asymptomatic. Humans, similarly, usually are not seriously harmed; a red bump forms, the lymph nodes might swell, and a mild fever may emerge. However, more serious infections can, and have, occurred, particularly among immune-compromised individuals. Although *Bartonella* is not as dangerous as plague, it is more common.

Cat scratches or bites can also lead to other diseases that are much more likely to be harmful. Such was the case with thirteen-year-old Grace Polhemus, of Brooklyn, New York. On October 18, 1913, while playing in her front yard, Grace bent over to pet a stray cat, and it bit her on her right wrist. It was discovered later, when brain tissue from the cat was tested, that the animal had rabies—another potentially fatal disease that can jump to humans from cats with a swipe of the paw or a quick nip of the mouth. As is often the case with rabies, the symptoms did not show up immediately in Grace. However, she died of the disease after slipping into a coma, a little over a year after being bitten.

The word *rabies*, taken from Latin, means "furious" or "to do violence." A highly infectious viral disease, rabies has likely been around throughout known human history. References to rabies from as early as the fifth century BC can be found in the writings of several prominent Greek and Roman scholars and philosophers, including Democritus, Aristotle, Hippocrates, and Virgil. Up until the late 1800s, a bite from a rabid dog (the primary way humans were exposed to rabies) was a death sentence. Whether by bite or scratch, once the virus enters the body, it travels along nerve fibers, jumping neuron to neuron, slowly making its way to the brain. All mammals are susceptible to rabies, and although dogs are still sources of infection for people in Asia, Africa, and India, in North America wild animals, such as bats, foxes, skunks, and Raccoons, are thought to serve as the main reservoirs of the virus. These species can transmit the disease to domesticated species like cats, cows, and horses, which can then become vectors of the virus. When humans are exposed to the rabies virus, symptoms usually appear within one to three months of exposure. If a post-exposure prophylaxis, or treatment, is not administered, one of two forms of the disease can develop. The most common form is "furious rabies," an early symptom of which is the fear of water—because of the difficulty swallowing—while the victim also experiences extreme thirst. For that reason rabies was also referred to as *hydrophobia*. Other symptoms might include hyperactivity and uncontrollable excited behavior (that is the "furious" part), extreme fever, tingling at the site of the bite, and eventually, as the virus spreads throughout the nervous system of the victim toward the brain, encephalitis (inflammation of the brain), and then death. The second type of rabies occurs in 30 percent of the cases and presents itself as a slow paralysis, usually starting at the site of the wound; patients eventually become comatose before dying. With both types, once symptoms appear death is almost certain. Fewer than ten people are known to have survived rabies infection since 1940; two of these people died within a few years of recovery from initial infection, and all but one had ongoing neurologic disorders.

Rabies is present on every continent but Antarctica. Despite the availability of highly effective pre-exposure vaccines (thanks initially to Louis Pasteur) as well as effective post-exposure prophylaxis, rabies is still responsible for more deaths per year worldwide than all other zoonotic diseases. The World Health Organization estimates that at least 60,000 people perish every year from rabies, primarily in Asia and Africa, and most of these are children under fifteen years of age. Stray dogs are still the primary reservoir and transmission mode of the disease, causing 90 percent of the exposures in Asia and Africa, and 99 percent of all human deaths. In 1946, prior to widespread vaccinations and control of stray dogs in the United States, there were 8,384 reported cases of rabid dogs and 455 cases of rabid cats. In 2010, thanks to effective and enforced policies to promote the vaccinations of owned dogs and the elimination of stray dogs, canine rabies had declined to only sixty-nine cases. The number of rabies cases in cats, however, has decreased at a less significant rate, to 303 in 2010. Since 1988 cats have been the number-one domesticated species passing rabies infections to humans. In 2013, 53 percent of all reported rabid domesticated species were cats, followed by dogs at 19 percent. The cause of this pattern seems clear—the presence of millions and millions of stray and unvaccinated free-ranging cats on the landscape, many of them sharing feeding stations with wildlife that are susceptible to rabies. Although bats, skunks, foxes, and Raccoons are the primary reservoirs of rabies in North America, cats, because of their high contact with humans, are the most important source of human exposure. The National Association of State Public Health Veterinarians does not mince words about stray cats and dogs, because of the risk of rabies and related serious health threats. The group's policy (2011) states:

> Stray dogs, cats, and ferrets should be removed from the community. Local health departments and animal control officials can enforce the removal of strays more effectively if owned animals are required to have identification and are confined or kept on leash.[1]

In fact, the Centers for Disease Control and Prevention and the Pennsylvania Department of Health specifically consider rabid cats to be a serious public-health concern. From 1982 to 2014 there were 1,078 laboratory-confirmed cases of rabies in outdoor domestic cats in the state of Pennsylvania (see fig. 5.1). This proliferation is likely linked to a rabies outbreak in Raccoons across the entire eastern seaboard that started in the 1950s. Outdoor cats regularly interact with Raccoons and other species when humans place food outside for wildlife and cats—especially in places like cat colonies. Such food abundance concentrates and focuses the potential for cat-wildlife interactions, providing increased opportunities for rabies transmission. Why aren't cats in colonies immune from rabies even though they are sometimes captured, vaccinated, spayed or neutered, and then released, as part of trap-neuter-return (TNR) programs? Unfortunately, a single vaccination is not sufficient. The American Veterinary Medical Association requires revaccinations within twelve months after the initial vaccination for all cats and even recommends a recurring booster for the most effective vaccination. Catching an unowned outdoor stray or colony cat is hard to do once; doing it a second time and getting the animal revaccinated may be about as likely as a major league baseball player throwing two no-hitters in a year. Thus, most unowned cats remain unprotected against the rabies virus. (There will be much more on the shortcomings of TNR programs later; see chapter 7.) Making matters worse, people, especially children (like Grace Polhemus), are much more likely to approach a cat than they are a "wild" species like a Raccoon, and cats can shed the rabies virus for several days prior to symptoms appearing. If an interaction occurs during this period, and rabies transmission is not suspected, it will be through the onset of rabies infection only—perhaps several weeks or months later—that a rabies transmission will be known to have occurred. At that point it is likely too late.

Thankfully, the development of rabies infections in humans due to cats and other animals is extremely rare in the United States. Only a few cases appear each year. The development of highly effective post-exposure prophylaxis procedures (along with the significant reduction of stray dogs) has been critical in preventing more human deaths. Nowadays, any time a human is suspected of being exposed to a rabid animal through a bite or scratch, a post-exposure prophylaxis procedure is administered. Although no standardized reporting is done for post-exposure prophylaxis applications across the United States, the vast majority of the 38,000 post-exposure rabies treatments conducted annually are the result of people interacting with a suspected rabid cat. Each of these post-exposure prophylaxis treatments costs public-health departments and U.S. taxpayers somewhere in the neighborhood of $5,000 to $8,000, amounting to at least $190 million across the United States each year.

🐾

These diseases, which result from spillover from outdoor domestic cats, constitute a serious public-health issue. Unfortunately, there are even more-insidious disease organisms that make cats their home. These organisms are more complex in fundamental structure and have more elaborate life cycles than the bacteria and viruses that spread plague and rabies. They jump to other species and then, through a complex series of changes, both physical and chemical, get the new host to change its behavior in a way that makes it easier for the parasite to continue to reproduce and transmit. The ability of these parasites to manipulate the behaviors of their host species makes these host-parasite relationships among the most fascinating case studies in the biological world. These are the organisms that are the inspirations for zombie movies, except they are not fictional.

Toxoplasma gondii is a single-celled protozoan parasite with a worldwide distribution. It is incredible in how it maneuvers through its primary host (e.g., felines) and how it ultimately, and quite destructively, manipulates the behavior of its secondary host (e.g., other animals, including humans) to further its transmission and existence.

Toxoplasma reproduces sexually only in the intestines of domestic cats and other felines, its definitive hosts. In felines it multiplies and sexually reproduces to produce oocysts—cysts containing the *Toxoplasma* zygote (the two-celled body formed from the fusion of sperm and egg through sexual reproduction). These oocysts ultimately are shed in copious quantities into the environment via cat feces, a process that occurs for several weeks after a cat is initially infected. The oocysts are extremely resilient once in the environment and can persist for months to years and under all kinds of conditions, including while submerged in fresh or salt water or in frozen soil. Secondary hosts, such as mice, rats, and birds, then either intentionally or accidentally ingest the cat poop infected with *Toxoplasma* oocysts or pick up the oocysts from the infected environment. (Humans, of course, can also pick up the oocysts or other forms of *Toxoplasma*—more on this later.)

Once in the secondary host, the *Toxoplasma* oocysts then transform into something called a tachyzoite and multiply asexually rapidly. Tachyzoites can be as small as one-tenth the size of red blood cells when they invade healthy cells. There they divide quickly, causing tissue destruction and spreading of the *Toxoplasma* infection to the new host organism. Eventually the infection localizes in muscle and nerve tissue—especially in parts of the brain—in the form of cysts called bradyzoites (fig. 5.2). Then something odd begins to happen to the newly parasitized host: its normal behavior of fear toward cats turns into attraction. Specifically, the smell of cat urine—a smell that uninfected mice and rats were thought to be hardwired to fear and avoid—becomes an attractive aphrodisiac. This is exactly how the *Toxoplasma* parasite wants its hosts to behave, because it turns infected rodents into easy prey. Once the infected host, along with the parasites infecting its body, is eaten by a new predator (preferentially a cat or other species of feline), the parasite can begin its sexual reproductive cycle again, infecting a new host, shedding oocysts, and expanding its reach.

Does the parasite *Toxoplasma* really manipulate the behavior of a secondary host organism, creating an almost fatal feline attraction, for its own benefit? Researchers at Oxford University believe the answer to that question is a resounding yes. Manuel Berdoy and colleagues experimentally tested the "parasite-manipulation hypothesis" by artificially infecting lab rats with *Toxoplasma* to determine whether the parasite interfered with the rat's innate reaction to predation risk by cats. The experiment consisted of examining the nocturnal exploratory behavior of twenty-three *Toxoplasma*-infected rats and comparing it to that of thirty-two uninfected rats. All rats appeared healthy regardless of infection status. The rats were placed in pens that had a layer of straw and a labyrinth of bricks creating a maze. Random corners of the maze were prepared with one of four treatments: the rat's own straw bedding, fresh straw bedding treated with water, straw bedding treated with cat urine, or straw bedding treated with rabbit urine. Rats were tested individually over an entire night. The results were striking. As would be expected, healthy uninfected rats exhibited a clear aversion to the areas of the maze treated with cat urine. In contrast, and consistent with the parasite-manipulation hypothesis, rats with *Toxoplasma* infection exhibited an attraction to areas with cat urine. Both types of rats behaved the same way in the presence of their own smell and that of rabbits. The results were consistent with the idea that the parasite somehow subtly manipulates the brain and thus the behavior of the rat. What remained unclear was how the *Toxoplasma* infection, the cysts themselves, disrupted the neural circuitry of the brain and whether the attraction effect was really specific to cats—a response that would confirm the parasite-manipulation hypothesis.

The neuroscientist Robert Sapolsky, the John A. and Cynthia Fry Gunn Professor at Stanford University, decided to explore those unanswered questions. His previous research examining how stress is perceived and then translated into actual chemical signals in the body, influencing brain chemistry and action, provided a useful foundation for this exploration.

Sapolsky, along with Ajai Vyas, Patrick House, and several other collaborators, led two research projects that demonstrated that *Toxoplasma* infection not only eliminated a rat's fear toward cat pheromones, it also stimulated an attraction by rats toward cats. The attraction was remarkably specific to cat urine. The group's research confirmed that the parasite did indeed manipulate a secondary host for its own benefit through a spectacular example of host-parasite evolution. Sapolsky and his colleagues were also able to demonstrate that post infection, *Toxoplasma* cysts settle disproportionately near the limbic regions of the brain—the parts of the brain that control instinct, mood, defensive behavior, and sexual attraction. *Toxoplasma* causes rats to become almost sexually attracted to cats—an attraction that proves fatal. At least ten peer-reviewed scientific studies have now been published confirming this finding.

Why does this matter to humans? Toxoplasmosis, the disease caused by *Toxoplasma gondii*, is one of the most common parasitic infection in humans. In fact, it is estimated that approximately 30 to 50 percent of the world's population and up to 22 percent of the U.S. population (more than 60 million Americans) are infected with *Toxoplasma gondii*—somewhat less than half of them due to direct ingestion of the *Toxoplasma* oocysts excreted into the environment by cats, most likely domestic cats. Many of the human infections are caused by eating infected and undercooked meat of domesticated animals that ingested the oocysts in food or water infected through cat poop. This form of toxoplasmosis lies quietly encysted in muscle and may be passed from carnivore to carnivore or omnivore (like people and pigs) and multiply in the food chain.

In some countries, human *Toxoplasma* infection rates are very high. Jaroslav Flegr, an evolutionary biologist at Charles University in Prague, Czech Republic, has been studying various aspects of *Toxoplasma* and toxoplasmosis for most of his career. He conducted a review of its prevalence in women of child-bearing age in eighty-eight countries and found that rates varied from a low of 4 percent in South Korea to highs of 84 percent, 78 percent, 63 percent, and 54 percent in Madagascar, Nigeria, Germany, and France, respectively. According to the Centers for Disease Control, humans in these various countries become infected with *Toxoplasma* oocysts from their environment in a variety of ways, including:

- Eating undercooked contaminated meat (especially pork, lamb, and venison) and/or accidental ingestion after handling contaminated meat and not washing hands (*Toxoplasma* cannot be absorbed through skin).
- Drinking water contaminated with *Toxoplasma*.
- Accidentally swallowing the parasite through contact with cat feces that contain *Toxoplasma*. This might happen by cleaning a litter box, touching anything that has come into contact with infected cat feces, or accidentally ingesting contaminated soil (e.g., from poorly washed fruits or vegetables).
- Mother-to-child congenital transmission.
- Organ transplant or blood transfusion.[2]

How likely is it that someone might ingest *Toxoplasma* oocysts, either directly or indirectly? In the United States alone 1.2 million metric tons of cat feces are defecated every year.

Despite the fact that cats are infectious and shedding oocysts for only about three weeks, oocysts are almost omnipresent. Research on the pervasiveness of *Toxoplasma* oocysts in California, France, Brazil, Panama, Poland, China, and Japan estimated a range of between three and 434 oocysts per square foot. Cats prefer to defecate on loose soil and often choose places like gardens and children's sandboxes if available. As a result, such places are known to have a much higher density of *Toxoplasma* oocysts. Because children under three years of age are known to put their hands in their mouths every two to three minutes and can ingest a measurable amount (up to forty milligrams) of soil per day, those playing in uncovered sandboxes are at extreme risk. Most of us have encountered that small candy bar–like object while playing with our

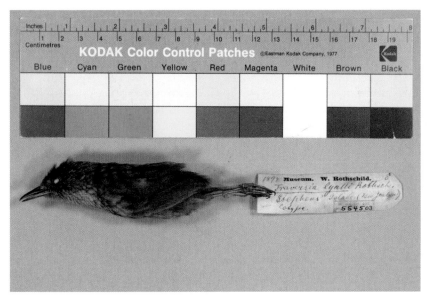

1.1 A Stephens Island Wren specimen prepared by David Lyall and acquired by Walter Rothschild, from the collection of the American Museum of Natural History. (Courtesy of AMNH staff photographer Matthew Shanley)

2.1 European Wildcat (*Felis silvestris silvestris*), a subspecies of the Wild-cat and close relative of today's domestic cat. (Courtesy of Alex Sliwa)

2.2 Stanley Temple holding tracking equipment and one of the cats in his 1989 study in the farmlands of Wisconsin. (Courtesy of University of Wisconsin–Madison Archives)

2.3 Rachel Carson and friends on a bird walk in Glover Archibald Park, Washington, DC, September 24, 1962. (Courtesy of Shirley A. Briggs)

2.4 A Socorro Dove (*Zenaida graysoni*), endemic to Isla Socorro off the southern tip of Baja California and last seen in the wild in 1972. (Elizabeth Barrett)

3.1 Roger Tory Peterson, painting one of the wader plates for *A Field Guide to the Birds of Britain and Europe*, while staying on Hilbre Island in the Cheshire Dee, England, in October 1952. Peterson's innovative field guides helped democratize the pastime of birding. (Courtesy of the Eric Hosking Charitable Trust)

3.2 Grumpy Cat, one of America's favorite Internet felines, at the 2014 MTV Movie Awards. (Jaguar PS/Shutterstock)

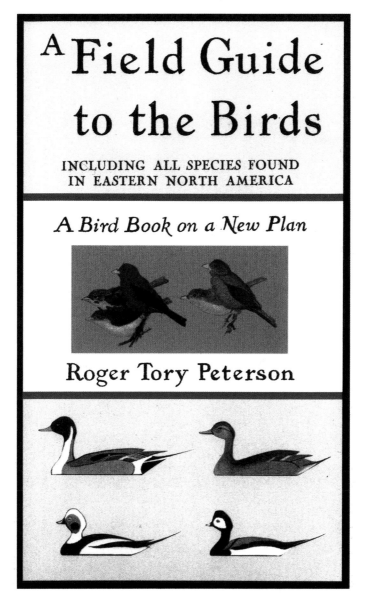

3.3 The cover of the first edition of Peterson's *Field Guide to the Birds*.
(Courtesy of Houghton Mifflin)

4.1 A cat with a freshly killed Hooded Warbler (*Setophaga citrina*). (Shutterstock)

4.2 A cat wearing a radio collar, used in a 2004 study by Roland Kays and Amielle DeWan investigating levels of predation by owned outdoor cats. (Courtesy of Roland Kays)

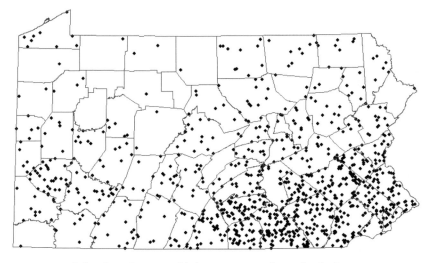

5.1 Map of the distribution of laboratory-confirmed rabid cats in Pennsylvania from 1982 to 2014. Rabies is one of many diseases that can spread from cats to people. (Courtesy of Leah Lind, PA Department of Health)

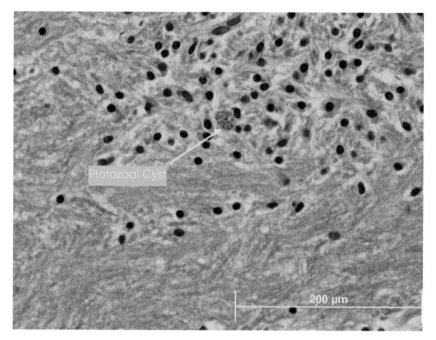

Protozoal Cyst

200 µm

5.2 Histology slide of a protozoal cyst, or bradyzoite, of the parasite *Toxoplasma gondii*, which has been linked to multiple illnesses in humans, including schizophrenia and bipolar disorder. (Courtesy of D. Rotstein and NOAA)

5.3 The endangered Hawaiian Monk Seal (*Monachus schauinslandi*) is vulnerable to toxoplasmosis. Inshore waters can become contaminated with the bacterium's oocysts from runoff from cat feces on land and can be potentially fatal to monk seals, Sea Otters, and other marine mammals. (Courtesy of M. Sullivan/NOAA)

6.1 The Piping Plover (*Charadrius melodus*) is a threatened shorebird that breeds in the northern Great Plains and along the Atlantic coast. Juveniles are especially vulnerable to predation right after hatching. (Courtesy of Frode Jacobsen)

6.2 Jim Stevenson (here holding a Western Diamondback Rattlesnake), an ornithologist based in Galveston, Texas, was at the center of a controversy concerning the killing of a free-ranging cat in order to protect nesting Piping Plovers in 2006. (Courtesy of Jim Stevenson)

6.3 Investor turned philanthropist/activist Gareth Morgan has spearheaded a number of campaigns to revitalize New Zealand's native species, including the Cats to Go campaign. (Courtesy of the Morgan Foundation)

7.1 Trap-neuter-return (TNR) advocates capture free-ranging cats in humane traps placed near colonies. After spaying or neutering, the animals are returned to the wild, where they can continue to prey on wildlife and spread disease. (Dave Zapotosky/*Toledo Blade*, July 2, 2013)

7.2 A group of free-ranging cats. Estimates place the number of unowned, free-ranging cats in the United States at 60 to 100 million. (Shutterstock)

7.3 A PETA demonstration against fur. It is interesting to note that the animal rights organization takes a strong position against TNR, seeing adoption or euthanasia as more humane options. (S. Bukley/Shutterstock)

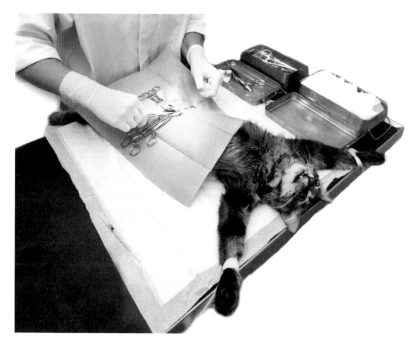

7.4 A spaying procedure under way. Once the cat is anesthetized, procedures take four to six minutes. The hard cost for the procedure at the Oregon Humane Society is $42.50. (Shutterstock)

7.5 While working at the San Francisco SPCA, Laura Gretch auctioned off squares of her skin to be tattooed with population control messages; the tattoo from the 2012 auction reads "Spaneuter." Auction proceeds went to the SPCA. (Courtesy of Laura Gretch)

8.1 "Catios" allow cats to enjoy the outside while limiting exposure to other animals and automobiles. (Catio Spaces)

9.1 Michael Soule, one of the founders of the field of conservation biology, with his indoor cat, Taz. (Peter P. Marra)

9.2 Two fabricated Stephens Island Wrens (*Xenicus lyalli*) in a setting reminiscent of their historical habitat on display in Te Papa Tongarewa (National Museum of New Zealand). (Courtesy of the National Museum of New Zealand, Te Papa Tongarewa)

kids in the sandbox or working in the garden. Then there is our drinking water.

After rains, runoff from human-dominated environments, often covered with hardscape, moves everything from pesticides to foam peanuts to encysted protozoans, including *Toxoplasma* oocysts, into freshwater and marine systems. These are the very water systems that meet the needs of millions of humans—through irrigating agricultural crops (carrots, potatoes, and lettuce), maintaining livestock, and filling reservoirs that entire cities use as sources for drinking water. *Toxoplasma* transmitted through such means can be quite harmful. This was the case in March 1995, in Victoria, British Columbia, where an outbreak of toxoplasmosis sickened at least 100 people. The origin was traced back to the municipal water system, although it was not determined whether the source of the infection was outdoor domestic cats or wild Cougars—since individuals of both species were found around the watershed and to be actively shedding the *Toxoplasma* virus. This was not an isolated case. Similar outbreaks have occurred with drinking water in Panama, India, and Brazil, all in part because of the oocysts' ability to withstand incredibly harsh conditions. Its persistence and impacts make *Toxoplasma* an environmental contaminant on the order of, if not worse than, DDT.

Only a single *Toxoplasma* oocyst needs to be consumed to result in an infection. Once the oocysts are ingested by a human, the tachyzoites divide rapidly in an acute phase of the disease. This can make a person quite sick—fever, fatigue, headaches. In people with compromised immune systems, such as late-stage HIV patients, even death can occur. It is unclear how diseases of the immune system (such as lupus, fibromyalgia, and chronic fatigue syndrome) and drugs used to suppress selected human immune function (like Cox-2 inhibitors) may affect latent *Toxoplasma* infections. Pregnant women and their fetuses have been known since the 1920s to be at serious risk. If infected with toxoplasmosis in the first trimester, one in ten fetuses will be aborted or become malformed—and this likely is an underreported statistic. Because of this problem, pregnant women have been warned for decades to avoid changing litter boxes and touching cat feces. Despite these warnings, congenital transfer of *Toxoplasma* continues to happen across the world.

In most people, a healthy immune system neutralizes active toxoplasmosis into what was thought to be a latent phase (but see below). Slowly asexually dividing bradyzoites form cysts in muscle and neural tissue (like the brain) and survive for the life of the human. The good news used to be that for the vast majority of humans living with latent-stage toxoplasmosis there were few quantifiable symptoms. Then scientists started to look a little deeper and found that bradyzoites were actually dynamic and replicating. In fact, one manifestation of toxoplasmosis infection is the development of ocular toxoplasmosis—basically cysts that settle in the eye. If the cysts burst they can cause a progressive and recurring inflammation of the retina that can result in glaucoma and eventually blindness. Regrettably, this is not the worst manifestation of toxoplasmosis infection in humans.

Thanks to recent research from Jaroslav Flegr and a handful of other pioneers (such as Manuel Berdoy, Jitender P. Dubey, Robert Sapolsky, E. Fuller Torrey, Joanne P. Webster, and Robert H. Yolken), it is becoming clearer and clearer that the latent phase of a toxoplasmosis infection is not as asymptomatic in humans as we once thought. There is now overwhelming evidence that the same behavioral changes associated with toxoplasmosis infection in rats and mice—reduced anxiety, less fearfulness, and an attraction to cat pee—are also seen in humans. Toxoplasmosis also changes human behavior—likely through changes in brain chemistry. Hundreds of studies are accumulating on the side effects of latent toxoplasmosis (detected by the presence of anti-*Toxoplasma* antibodies in the blood). In addition to the changes in behavior similar to those seen in rodents, individuals with latent toxoplasmosis are showing symptoms that suggest a much larger array of mental illnesses, including severe depression, bipolar disorder, obsessive-compulsive disorder, and schizophrenia. One recent study found that across twenty European countries, suicide rates in older postmenopausal women were significantly and positively associated with rates of *Toxoplasma* exposure. A study in Denmark enrolled a cohort of 45,788 women who gave birth to their first child between 1992 and

1995 and followed them until 2006. All the women had their levels of *Toxoplasma* antibodies measured. Consistent with the broader study of European women mentioned above, those infected with toxoplasmosis were two times as likely to commit suicide than the women without toxoplasmosis infection. Flegr believes that collectively toxoplasmosis, either through the acute stage of infection or through mental and neurotic illness manifested during the latent phase, has contributed to the deaths of hundreds of thousands of people, if not significantly more, over the last few decades.

Schizophrenia is a severe brain disorder in which people interpret reality abnormally, resulting in some combination of hallucinations, delusions, and extremely disordered thinking and behavior. About 1.1 percent of American adults (2.5 million individuals) live with schizophrenia; the associated costs total between $40 and $60 billion per year. Psychiatrist Fuller Torrey, the executive director of the Stanley Medical Research Institute in Chevy Chase, Maryland, has been studying schizophrenia for almost his entire career. He has written twenty books and published more than 200 papers, many of which are on the topic of schizophrenia. Torrey and neurovirologist Robert Yolken (the director of the Stanley Division of Developmental Neurovirology at Johns Hopkins University, in Baltimore) have been collaborating to understand how infectious agents, such as *Toxoplasma*, might be responsible for the onset of schizophrenia. In one of their papers on the topic, they conducted a review of almost fifty independent papers that examined the relationship between the presence of *Toxoplasma* antibodies and schizophrenia. In their meta-analysis (a statistical technique of combining the results of multiple studies into one to determine whether there is a general or overwhelming effect) they found that individuals infected with *Toxoplasma* were 2.7 times more likely to develop schizophrenia compared to individuals without *Toxoplasma* infection. Furthermore, four recent studies report that individuals with schizophrenia, compared to controls, have had more contact with cats during childhood. Today Fuller Torrey publicly states that although permanent indoor cats are relatively safe, he would not have a cat, especially a kitten, in contact with a child if the cat is going outside. He would be concerned about the risk of that child developing schizophrenia later in life. After more than twenty years of research on toxoplasmosis and various mental illnesses like schizophrenia, Torrey believes that *Toxoplasma* oocysts pose a significant public-health hazard. Jaroslav Flegr agrees and believes that malaria, now considered to be the most devastating protozoan killer of humans, will be "dethroned" by toxoplasmosis. As long as we continue having outdoor cats, the parasite will spread. Once humans are infected and symptomatic, most will recover. Some, however, will need to be treated with a combination of drugs, perhaps for the rest of their lives. The parasite is not likely to be eliminated, due to the locations where it encysts. It is becoming increasingly clear that toxoplasmosis is a significant zoonotic disease—perhaps one of the most significant—that impacts humans globally, and it emerges primarily from outdoor domestic cats.

Toxoplasmosis is also a significant killer of wildlife, including some of the most endangered species on the planet. When *Toxoplasma* oocysts contaminate our land and then pollute our fresh and marine waters through runoff, they get into food webs and eventually cause mortality in a variety of marine mammals and birds.

❖

There are three species of monk seals on earth, all of which live in tropical marine environments. One of these species, the Caribbean Monk Seal (*Monachus tropicalis*), is considered extinct. Populations of another, the Mediterranean Monk Seal (*Monachus monachus*), are thought to hover at around 500 individuals. The third, also among the most imperiled marine mammals on the planet, is the endangered Hawaiian Monk Seal (*Monachus schauinslandi*; fig. 5.3). Fewer than 1,000 individuals are thought to remain, and their populations have been declining at a rate of about 10 percent a year since 1989. Distributed throughout the Northwestern Islands and the main Hawaiian Islands, Monk Seals face numerous threats, including entanglement with marine debris, food limitation, and—we now know—toxoplasmosis. Cats are in tremendous abundance all over the islands of Hawaii, and *Toxoplasma* has been present and circulating since at least the 1950s. When it rains in Hawaii, which it does a lot, cat feces and the *Toxoplasma* oocysts it contains are carried into nearshore marine waters. A minimum of eight Hawaiian Monk Seals have been found dead due to toxoplasmosis in the last ten years (two just in 2015), and this likely is a significant underestimate. The National Oceanic and Atmospheric Administration (NOAA), the branch of the U.S. federal government responsible for the management and protection of Hawaiian Monk Seals, now views free-ranging cats and the *Toxoplasma* oocysts they spread as a serious threat to the dwindling seals. It has become clear that cats not only kill native species directly through predation, they also do so indirectly through the shedding of *Toxoplasma* oocysts.

The 'Alalā, or Hawaiian Crow (*Corvus hawaiiensis*), is another example, and another species endemic to Hawaii. The last two wild individuals were seen in 2002, and the species is now extinct in the wild. Thankfully, a captive breeding program had been established, and today more than 100 Hawaiian Crows are alive in captivity. The species' initial declines were thought to have been caused by predation by rats, mongooses, and cats, as well as habitat destruction and introduced diseases like toxoplasmosis. Understanding the impacts of disease on wild birds, much as with quantifying the impacts of predators like cats, is extremely difficult to do in the field. In the 1990s, in an effort to restock populations in the wild, scientists fit twenty-seven Hawaiian Crows with radio transmitters and released them into the wild. At least five of these birds became sick with toxoplasmosis infection. One was recaptured and treated and slowly recuperated. The other four individuals were found dead in the field and were determined to have died from toxoplasmosis. Given the sensitivity of the Hawaiian Crow to toxoplasmosis, any future reintroduction efforts will need to consider the impact of free-ranging cats and *Toxoplasma*.

The list of species impacted by *Toxoplasma* infection is long, but the parasite's predominance among marine mammals has caught scientists by surprise. The catalog of *Toxoplasma* casualties in marine mammals includes seals, sea lions, dolphins, Antillean Manatees, Beluga whales, and Sea Otters. And more victims are likely yet to be discovered. The ability of *Toxoplasma* oocysts to survive in marine environments and to move from terrestrial to marine ecosystems and up in the food chain to top-level consumers demonstrates the resiliency of these oocysts and the interconnected nature of these ecosystems.

Among marine mammals impacted by toxoplasmosis, the Sea Otter (*Enhydra lutris*) is one species that has been struggling to persist on the planet for a number of years and is now on the U.S. endangered species list. Sea Otters were almost wiped out in the early twentieth century, when their numbers plummeted to between 1,000 and 2,000 individuals. And while they have slowly clawed their way back in some parts of the West Coast of the United States from California to Alaska, they continue to decline in others. Hunting, oil spills, and marine pollution have been obvious culprits in Sea Otter demise. *Toxoplasma* infections have more recently hit the stage. A group from the University of California, Davis, and the California Fish and Game Commission trying to suss out causes of Sea Otter deaths conducted postmortem analyses of 105 dead Sea Otters collected along the California coast from 1998 to 2001. *Toxoplasma* infection and shark attacks were the two leading causes of mortality, and these likely were linked. Individuals with fatal shark bites were over three times more likely than those that died of other causes to have preexisting *Toxoplasma* infections. Given what we know about how *Toxoplasma* changes the behavior of rodents, we can imagine it possible that *Toxoplasma*-infected Sea Otters also became less fearful of their perennial predators, or at least became sick and therefore unable to escape them. Regardless, *Toxoplasma gondii* is a significant threat to endangered Sea Otters, Hawaiian Monk Seals, and numerous other species—including, of course, humans.

‸

Unfortunately, there are also other deadly pathogens carried by domestic cats, including feline leukemia. This pathogen can impact both domestic and wild felines, and can be transferred from domestic animals to native species.

Feline leukemia is found in domestic cats worldwide. In the United States, 2 to 3 percent of all cats are infected with this virus, although these rates can go as high as 47.5 percent in colony cats and vary depending upon the cats' age, sex, and condition. Once a cat is infected, feline leukemia virus can be deadly. Infected cats easily spread the virus through saliva, nasal secretions, urine, and feces. Feline leukemia virus makes cats very sick and is the major cause of cancers in domestic cats. The virus is also spread if an infected cat happens to be consumed.

One endangered feline impacted by the feline leukemia virus pathogen is the Florida Panther (*Puma concolor coryi*), a subspecies of the Cougar. Its populations, which once occurred throughout the southeastern United States, were decimated starting in the 1600s with massive land clearing. By the 1970s the panther had been isolated in southern Florida and was at the brink of extinction as its population size dropped to a low of twenty individuals. Today, thanks to land protection in Florida and the introduction of Cougars from Texas to reduce inbreeding, the population has grown to between 100 and 160 animals. Florida Panthers are by no means out of hot water. Other threats continue to loom, such as injuries sustained from fights with fellow panthers and collisions with cars. The consequences of small population size (as described in chapter 4) such as genetic malformations and compromised immune systems make them especially vulnerable to disease. In 2002 an outbreak of feline leukemia virus, originating from free-ranging domestic cats, reached the Sunshine State and went on to kill at least five endangered panthers through 2005. Wildlife biologists and veterinarians worked diligently to capture and inoculate as many uninfected panthers as possible with a new vaccine developed to protect them from the virus.

Two other native cats in the United States have been known to contract and die from feline leukemia virus: the Cougar subspecies of the western United States (*Puma concolor couguar*) and the Bobcat (*Lynx rufus*) throughout the country. Both of these wild felines are known to consume free-ranging cats and are likely to succumb to the disease if infected. Cat species native to other continents have also been impacted by feline leukemia virus, including the European Wildcat in Scotland, Spain, and France; the endangered Iberian Lynx in Spain; the Puma (Cougar), Ocelot, and Little Spotted Cat in Brazil; and the list goes on.

Free-ranging cats clearly pose a significant threat to a number of wild animals—birds, small mammals, reptiles—that are vulnerable to their predation. Some of the birds and animals they kill totter on the brink of extinction. However, the bacteria, viruses, and parasites cats carry and transmit—*Yersinia pestis* (plague), rabies virus, and *Toxoplasma gondii*, among others—also can drive wildlife species to extinction. Additionally, cat-transmitted pathogens have impacted millions of humans and pose one of the least understood but most critical public-health challenges of our time. Surely, some action must be taken to lessen free-ranging cats' impact on animals and people. One solution is to remove them—once and for all—from the landscape.

CHAPTER SIX

Taking Aim at the Problem

The man of science has learned to believe in justifi-
cation, not by faith but by verification.
—Thomas Huxley

Where water meets land a unique fringe habitat emerges. Mud-
flats, marshes, and shorelines allow species of amphipods, cope-
pods, worms, clams, and crabs to burrow, skitter, and spawn. More
than 200 species of shorebirds have evolved different length legs
and bills, uniquely adapted for removing these precious protein resources (either the organisms themselves or the eggs they leave behind) from different depths of mud and water. In the United States
examples include the Long-billed Curlew, Marbled Godwit, Willet,
Greater Yellowlegs, and Least Sandpiper. One family of shorebirds,
known as the plovers, is represented by sixty-six species globally.
While most shorebird species breed in the far northern reaches of
the globe, often on subarctic and arctic tundra, a handful breed in
coastal areas of temperate regions, including one species, the Piping Plover (*Charadrius melodus*; fig. 6.1).

In Roger Tory Peterson's *Field Guide to the Birds*, the Piping
Plover is distinguished from other plovers in the summer by its
pale back (the color of dry sand) and its very short, stubby bill
and orange legs (which are brightest in breeding season). Small
(roughly the size of a sparrow) and muted in color, the Piping

Plover is the kind of bird that might be overlooked by the casual observer or mistaken for a sandpiper or other small shorebird. In the summer it is found along the dunes, sand flats, and beaches of the Atlantic coast from the Canadian Maritimes to North Carolina. Populations are also found around the shores of the Great Lakes and along rivers, lakes, and wetlands of the northern Great Plains. In the winter birds from these different breeding populations congregate on beaches and barrier islands, preferring open sandy stretches or rocky shores away from the water, along the southern Atlantic coast from North Carolina to Florida, at a few places in the Caribbean, and along the Gulf Coast as far south as the Yucatán. Piping Plovers capture prey, including marine worms, insects, mollusks, and crustaceans, at the land-water fringe. They forage using a run-stop-scan technique, leaning forward and picking at surfaces when they spot their quarry. Plovers also employ a "foot-tremble" feeding method, causing prey to move and thus become more conspicuous. Their eponymous call—a plaintive, bell-like whistle—is often heard before the birds are visible.

Piping Plovers reproduce in the early spring immediately upon arriving on breeding grounds in the North. Most of the birds nest in coastal areas, laying three to four eggs in a shallow depression lined with pebbles and fragments of shells. Eggs hatch within thirty days, and young plovers can fly within thirty days of hatching. Though they blend in well with their surroundings, Piping Plover eggs and chicks are especially vulnerable to storms and abnormally high tides, as well as predators—foxes, Raccoons, crows, and cats, among them. When predators do appear, adults will frequently feign a "broken-wing" display to distract them away from the nest. Adult Piping Plovers are also sensitive to human presence. If their nesting sites are too frequently disturbed by beachgoers—be they four-wheelers, kite flyers, or picnickers—they will abandon their eggs. Plovers begin their southward migration in August and spend the remainder of the fall and most of the winter months on their nonbreeding grounds in the southern United States, the Bahamas, and the Caribbean before migrating north once again the following spring.

Development on coastal fringe habitats began to accelerate after World War II and has put Piping Plover populations in steep decline. Today biologists estimate that there are roughly 8,000 adult birds remaining. The birds were listed as threatened and endangered on January 10, 1986, in accordance with the U.S. Endangered Species Act (ESA) of 1973. The goal of such a listing is to protect the species to allow its populations to rebound to a healthy enough level that it can eventually be delisted. The Piping Plovers that nest in the Great Lakes area are classified as endangered (the most critical status), while the northern Great Plains and Atlantic coast birds are classified as threatened (or near-endangered). All the birds are treated as endangered during the winter nonbreeding season, since their breeding origins are not confidently known. The U.S. Fish and Wildlife Service identifies two primary reasons for the plover's endangered status: habitat loss and degradation (due to coastal habitat development and changes in water levels due to dams and other water control structures) and nest disturbance and predation (from the presence of both humans and predators near nesting sites).

The United States defines endangered species as animals and plants that are in danger of going extinct, and threatened species as animals and plants that are likely to become endangered in the foreseeable future. Under the ESA, Congress took action to provide for "the conservation of ecosystems upon which threatened and endangered species of fish, wildlife and plants depend."[1] The act prohibits, among other activities, "takings" of listed species—interpreted to include actions to "harass, harm, pursue, hunt, shoot, wound, kill, trap, capture, or collect, or to attempt to engage in any such conduct."[2]

🐾

Jim Stevenson did not have the language of the Endangered Species Act specifically in mind when he dropped a loaded .22 caliber rifle into his white Dodge van on the morning of November 8, 2006 (fig. 6.2). But he was hell-bent on preventing anymore "takes" of Piping Plovers perpetrated by a colony of feral cats living under the bridge at San Luis Pass, a channel that connects Galveston Island to Follets Island on the Gulf of Mexico southeast of Houston.

Stevenson, a stocky man then in his mid-fifties, had been a high school science teacher before founding the Galveston Ornithological Society. He had been out to the pass the evening before to do some bird-watching. This region of the Gulf Coast is celebrated among birders for its abundance of shorebirds and for the hundreds of millions of Neotropical migratory birds that stop over here to refuel and rest in the early spring during their return to northern climes. That evening Stevenson had spied a small group of Piping Plovers in the dunes near the bridge—and a cat stalking the birds. The presence of the cat infuriated Stevenson, who knew the plovers were endangered and needed protection. Here was a free-ranging cat, an invasive species, left unmolested to make a mockery of the efforts to restore Piping Plover populations to what the ESA deems a "healthy and vital" state. When he reached the pass that morning, Stevenson could make out the cat that had been stalking the plovers the previous evening among some other cats under the bridge. He took aim with his .22, and soon the cat lay dead. Stevenson's eco-vigilantism did not go unnoticed. John Newland, a tollbooth attendant on the road above, heard the shot and saw Stevenson's van pulling away. Newland had been a caretaker of the cat colony, providing food and water for the animals on a regular basis. He had become attached to the cats. As irate about the shooting as Stevenson had been with the cat's hunting, Newland quickly alerted the local police, and soon Stevenson was sitting in jail. He was accused of animal cruelty, an offense in the state of Texas that at the time carried a maximum penalty of two years in prison and a fine of $10,000. According to Section 42.09(2) of the Texas code, "Cruelty to Nonlivestock Animals," a person commits an offense of cruelty if he or she "intentionally or knowingly":

1. tortures an animal or in a cruel manner kills or causes serious bodily injury to an animal;
2. without the owner's effective consent, kills, administers poison to, or causes serious bodily injury to an animal;
3. fails unreasonably to provide necessary food, water, care, or shelter for an animal in the person's custody;

4. abandons unreasonably an animal in the person's custody;
5. transports or confines an animal in a cruel manner;
6. without the owner's effective consent, causes bodily injury to an animal;
7. causes one animal to fight with another animal, if either animal is not a dog;
8. uses a live animal as a lure in dog race training or in dog coursing on a racetrack; or
9. seriously overworks an animal.[3]

Stevenson was clearly holding the smoking gun, though it was possible the law was on his side.

Stevenson's fate depended, at least in part, on how the state of Texas would define the legal status of the cats that Newland had been feeding. Were they pets? Or pests? This was no simple matter. The Animal Welfare Act (U.S. Code section 2131) provides for the humane care and treatment of pets. And many states have codified laws concerning animal cruelty, vaccination requirements, and penalties for abandonment of pets. Still, while most states have a number of regulations concerning dogs, there are very few "cat codes" on the books—including codes that indicate whether unowned free-ranging cats are pets or pests. As of this writing, cats are required to be licensed under state law only in Rhode Island; in several states, including Virginia and Louisiana, local governments have taken the initiative to create their own cat-licensing protocols.

Free-ranging cats occupy an ambiguous space on the legal continuum, not quite pet yet not quite pest. "A feral animal by legal definition is an animal who was domesticated but has escaped and lives in the wild without human support. They are not considered wildlife and therefore not within the control of state Fish and Game departments," explained David Favre, the Nancy Heathcote Professor of Property and Animal Law at Michigan State College of Law.[4] State or local governments may assert jurisdiction for purposes of adopting laws if they want to, but there is nothing to mandate that they act one way or the other. Legal confusion arises, according to Professor Favre, because different localities have adopted different public-policy approaches depending on the political pressures and information available as the decisions were made. "It is not the legal status that is causing the problem; it is that there is not political consensus about what to do," Professor Favre added.

Two areas of federal law that could have come to bear on Jim Stevenson's case are the Migratory Bird Treaty Act (MBTA) of 1916 and the Endangered Species Act. According to the MBTA, any person (or entity, such as a corporation) convicted of violating the act can be fined up to $15,000, imprisoned up to six months, or face both penalties. Corporations have been held accountable for violating the MBTA in incidents ranging from the leaking of a company's pesticide into a pond (where it posed a danger to birds) to the improper insulation of a utility's power lines (resulting in the electrocution of birds). Pamela Jo Hatley, a land-use attorney in Florida, has raised the question of whether a person violates the MBTA when he or she releases a cat into the wild and that cat kills a migratory bird. If an entity can be held responsible for an accidental chemical leak that results in the death of a migratory bird, why can't an individual be held responsible if the death of a migratory bird can be attributed to a cat or cat colony that can be traced to a negligent pet owner or an individual or organization that promotes colony care?

At the time of Stevenson's arrest, the Endangered Species Act had already been successfully applied to prosecute such indirect "taking" of endangered birds in Hawaii, as Hatley points out, in the case of *Palila* v. *Hawaii Department of Land and Natural Resources*. The state was maintaining feral sheep and goats on public land for hunting purposes, and these animals—which, like cats, are incidentally invasive species—were eating a native tree species, Māmane (*Sophora chrysophylla*). These trees provide both food and shelter for the Palila (*Loxioides bailleui*), a species of Hawaiian honeycreeper whose distribution is now restricted to the upper slopes of the Mauna Kea volcano on the island of Hawaii. The Ninth Circuit Court of Appeals held that the destruction of critical habitat upon which an endangered species depends harms the species. As the Hawaii Department of Land and Natural Resources was responsible for the presence of the sheep and goats, it was held liable. Since the ruling, in order to rid the habitat of the invasive sheep, the state has made the sheep-hunting season on Mauna Kea year-round and has eliminated bag limits. Increased recreational hunting has been supplemented with aerial shooting conducted by the Department of Land and Natural Resources as well as with animal drives. Though the Palila have shown modest signs of recovery, many sheep remain. Three thousand animals were removed from the volcano in 2013 alone. Hatley notes that in Florida some local governments have adopted ordinances that authorize the maintenance of cat colonies. In this scenario, these municipalities could find themselves liable under the ESA should these colonies result in the take by feral cats of an endangered species. It is not much of a stretch to surmise that the caretaker of a colony of feral cats could be held responsible for the killing of endangered species conducted by his or her colony members. This was precisely the defense that Stevenson's attorney—Tad Nelson—planned to deploy.

The colony of free-ranging cats seeking shelter under the bridge at San Luis Pass in Galveston (as of this writing at least three cats remain) is just one of tens of thousands of such outside-cat colonies. Myriad factors have led to the explosion in the number of free-ranging cats wandering America's urban, suburban, and rural landscapes. Cat abandonment by irresponsible pet owners is a primary factor, and the growing contingent of "colony caretakers" and other individuals providing varying levels of feeding and care for cats is another. The latter group has helped extend the life span of outside felines. Another factor is the trend in animal-management circles such as humane societies and animal shelters toward "no-kill" policies. Not so long ago Jim Stevenson may not have felt compelled to visit San Luis Pass; an animal control officer may very well have been out there to remove and eventually euthanize the cats—saving the Piping Plovers from the fate of becoming cat prey. And few people would have batted an eye.

🐾

In 2005, at the spring meeting of the Wisconsin Conservation Congress in La Crosse, a small city in the southwestern corner of the state near its border with Iowa and Minnesota, a fireman named Mark Smith stepped to the lectern. Smith proposed to the meeting's attendees that farmers, hunters, and other residents be allowed to kill stray domestic cats in order to control their population; at the time, all domestic cats were on the state's protected species list. He was frustrated with the free-ranging cats that congregated around his backyard bird feeder, opportunistically waiting to prey on avian visitors. Smith claimed that he was not a cat hater, but simply wanted people to keep cats under control. The resolution passed fifty-three to one at the La Crosse hearing, meaning that it would be placed on the agenda for the 2005 Wisconsin Conservation Congress. The WCC is an independent organization created by the state to gather public input on conservation issues. Its recommendations are not binding but are passed along to the Wisconsin Department of Natural Resources for its consideration.

This was not the first time the idea of declaring an "open season" on free-ranging cats had been put forth by a member of the public to the WCC. The previous effort, floated in 1999, was voted down by the congress before it could garner much attention. Employees of the Wisconsin Department of Natural Resources (among other state agencies) almost certainly wished this had been the fate of what came to be known as Q62.

"Question 62–Feral Cats," as the proposal was named, read as follows in the pamphlet available to the voting public:

> Studies have been done in Wisconsin [a reference to Temple and Coleman; see chapter 2] concerning effects of free-ranging feral domestic cats. These studies showed free-ranging feral domestic cats killed millions of small mammals, song and game birds. Estimates range from a minimum of 47 million up to 139 million of songbirds killed each year. Free-ranging feral domestic cats are not a native species in Wisconsin. The above mentioned cats do however kill native species therefore reducing native species.

At present free-ranging feral domestic cats are not defined as a protected or unprotected species. Thus Wisconsin should move to define free-ranging feral domestic cats, as any domestic type cat which is not under the owner's direct control, or whose owner has not placed a collar on such cat showing it to be their property. All such defined free-ranging feral domestic cats shall be listed as an unprotected species. In so doing Wisconsin would be defining and listing free-ranging feral domestic cats.

62. Do you favor the DNR take steps to define free-ranging feral domestic cats by the previously mentioned definition and list free-ranging domestic feral cats as an unprotected species?[5]

The WCC holds public meetings simultaneously in all seventy-two counties across the state on the second Monday in April to allow citizens to comment and provide their input on proposed fish and wildlife rule changes. That year's meeting was slated for April 11. Outside-cat advocates began rallying support to oppose the resolution while the Badger State was still cloaked in deep midwinter cold. Ted O'Donnell, the owner of a Madison-area pet store called Mad Cat formed a group called the Wisconsin Cat-Action Team (Wisconsin CAT) and launched a website called DontShoottheCat.com on February 16. The media attention this grassroots effort attracted underscores the fervor the issue arouses, especially among cat advocates. A few days later a local alternative news outlet picked up the story, and from there it spread farther and farther afield, being picked up by Reuters, Associated Press, and Fox News. O'Donnell ultimately landed on ABC's *World News Tonight*, the network's flagship news program. After he is introduced, O'Donnell, sporting a "Don't Shoot the Cat" T-shirt, declaims that hunters shooting cats will damage Wisconsin's progressive reputation and tourism industry. During another interview, on CNN, O'Donnell questioned the validity of the Stanley Temple–John Coleman report on the impact of cat predation on Wisconsin's wild birds (discussed in chapter 2), which had been cited by the state Department of Natural Resources in the write-up concerning the resolution. O'Donnell even suggested that its results had been skewed because of the "relationship between Temple and the American Bird Conservancy, which is probably the most rabidly anti-cat special interest lobby group in the United States."[6] For the record, American Bird Conservancy states that it is "dedicated to achieving conservation results for birds of the Americas." The fireworks surrounding the discussion of Q62 are wonderfully documented in *Here, Kitty Kitty*, a film by Andy Beversdorf that uses the controversy spawned by Q62 to explore the larger question of what to do about free-ranging cats. In the film, a representative for the Wisconsin Department of Natural Resources reports that she personally responded to 2,000 phone calls and 5,000 e-mails during the buildup to the April 11 vote. Both Mark Smith and Stanley Temple reported receiving death threats. (Temple's study from a decade earlier stirred up the vitriol directed at him.) In the film, Temple plays back a chilling message from his office phone: A female voice snarls, "You cat-murdering bastard, what goes around comes around. I declare Stanley Temple season open." He also reports that cars would cruise into his driveway in the middle of the night and that he showed up at his office some mornings to find threatening notes tacked on his door. People who knew Stan would say later that it was the only time they ever recalled seeing him visibly shaken. For the record: Stanley Temple has never killed or proposed killing a cat; he also had nothing to do with the emergence of Q62. Temple reported that the coauthor of the report, his graduate student John Coleman, was so disturbed by the many nasty and sometimes frightening threats from fanatics that he simply wanted nothing to do with cats after successfully defending his PhD.

At one point the film captures public comments at the April 11 Wisconsin Conservation Congress meeting in Madison. As more than one news outlet reported at the time, the fur did indeed fly in the packed room at the Alliant Energy Center. Some attendees wear kitty ears and whiskers; others are decked out in camouflage hunting gear. Moderators struggle to maintain order and civility as concerned citizens offer their input. One woman steps to the microphone and explains that she owns sixty-five feral cats and has spent hundreds of thousands of dollars on their care over the years, as onlookers gasp in a mix of wonder and horror. A man states that some advocates for the new regulation have emphasized the fact that cats are not native to Wisconsin and points out that neither are white people—to great laughter. At the end of the tumultuous evening, votes were cast in Madison and in the other seventy-one counties around the state: 6,830 votes were cast in favor of Q62, 5,201 in opposition.

Although a majority of WCC attendees—57 percent—had favored a change in the Wisconsin regulations to allow for the hunting of "'free-ranging' domestic feral cats," the notion proved a nonstarter. On May 17, 2005, the Wisconsin Conservation Congress announced that the measure would not be pursued any further. Despite the fact that WCC members had voted to present the measure to the Wisconsin Natural Resources Board, the executive committee declined to do so. To become law, the board would have needed to approve the measure for consideration by the Wisconsin legislature, which in turn would have had to pass the measure. Even after that, it would have become law only upon receiving the governor's signature.

On a practical level, even if Q62 had become law, it would have likely had little impact on Wisconsin's swelling free-ranging cat population. Cat hunting is currently legal in Minnesota and South Dakota to the west and has had little impact on the free-ranging feline populations there. "We don't make much of a dent in mice, rats, pigeons, starlings, or sparrows," Temple notes toward the end of the film, implying that listing a species as an unprotected animal (as those species are) is an ineffective way of controlling their numbers. And in rural areas it was already fairly common practice to accord unwanted free-ranging cats a sudden and violent end; "shoot, shovel, and shut up" was the mantra in such regions. There was little reason to think that the hubbub concerning Q62 would alter this behavior, though several cases concerning rural cat slayings did receive coverage in Wisconsin in the months thereafter, including that of Myrtle Maly, dubbed the "Septuagenarian Cat Killer." Maly admitted to resorting to poison after her neighbor's cats repeatedly entered her yard and attacked birds, and her calls to animal control yielded no assistance.

Ten years after the fact, what does Wisconsin's Q62 imbroglio say about American attitudes toward cats and birds? Given the Wisconsin Conservation Congress Executive Committee's unwillingness to pass on the WCC's recommendation to the Natural Resources Board—despite a decided citizen majority of supporters—one can conclude that political bodies have little stomach for taking on the cat lobby. Indeed, then-governor Jim Doyle went on record before the meeting of April 11, 2005, to say that should a feral cat hunting bill reach his desk, he would refuse to sign it. Prominent conservation groups—notably the National Audubon Society—also seemed unwilling to take a strong position on the issue, most likely fearing alienation from a portion of their membership. Many bird lovers, after all, also keep cats.

"Nothing has changed at all in terms of policy," Stan Temple said recently, when asked to reflect on Q62. "There's been no response from federal agencies in terms of articulating a policy for dealing with free-ranging cats. The greatest lesson I took from Q62 was to not allow the media to take control of the message."[7] In most of the media stories, there was no countering of any of the claims that cat advocates shared concerning cat behavior and impacts. American authorities remain resistant (if not heartily opposed) to the idea of managing free-ranging cat populations by lethal means.

🐾

In Australia a very different ethos is guiding free-range cat management strategies, as government officials rally to save a host of endangered endemic species from extinction.

Domestic cats arrived on the Australian Continent with European visitors, perhaps as early as the seventeenth century (with shipwrecked Dutch sailors) and most certainly by the late eighteenth century when the English began their colonization efforts. (Australia, like Antarctica, does not have any native members of the cat family, Felidae.) By the mid-nineteenth century, feral cat colonies were well established throughout most of the continent, with the exception of wetland rain forests and some offshore islands. Additional cats were intentionally introduced on the continent in the late 1800s in hopes of reducing populations of nonnative rabbits, rats, and mice.

Cats, as has been noted, are highly effective predators, and in the course of several hundred years they have had a significant impact on Australia's indigenous fauna and, ironically, no impact on its populations of nonnative rabbits, rats, and mice. In fact, a number of small mammals (the Australia Department of the Environment places the number at twenty-seven) and several species of ground-dwelling birds have gone extinct, thanks in large part to cat predation. Predation by foxes, another introduced species, has undoubtedly also contributed to the problem. "Many Australian mammals were easy prey for feral cats and foxes," said Dr. John Woinarski, a professor at the Research Institute for the Environment and Livelihoods at Charles Darwin University, who has been involved in research, management, advocacy, and policy relating to biodiversity conservation, particularly in relation to threatened species. "Before their introduction, there were no comparable native predators. All the mammals that went extinct were small, shy, nocturnal creatures like the desert bandicoot, a small rodent-like animal that was distributed through central Australia's arid regions. People didn't really appreciate them, as they weren't very aware of them. It didn't help that these animals had low reproductive rates."[8]

The connection between feral cats and these Aussie extinctions has been made most forcefully by Dr. John Wamsley, known in some circles as the "cat hat man." Beginning in the early 1970s, Wamsley would show up at public events wearing a hat fashioned from the pelt of a feral cat and fronted by the animal's face. He recalled in a 2005 interview that some animal liberationists had pointed out that it was illegal for him to do anything about the feral cats that were destroying wildlife on his land, and they would take action against him if he did so—so he had to change the law. His cat hat statement certainly attracted attention; in the same interview, he recalled, "I knew exactly what newspapers had reported it, by where the death threats were coming from."[9]

"Wamsley is a bit of a crank in some ways, but he's certainly charismatic," Woinarski said. And Wamsley's commitment, Woinarski acknowledged, has gone beyond controversial headgear. "He built some enclosures designed to keep cats and foxes out so native animals could live without these invasive predators. These experiments in creating a cat-free environment showed that the native animals could thrive without predators." The cat hat man's antics and experiments in predator-free environments certainly raised awareness of the impact cats had been having on Australia's native fauna. An analysis undertaken by the Australia Department of the Environment of the factors that had led to the extinction of the twenty-seven endemic mammals implicated cats as one cause, further strengthening Australia's resolve. "Once the report came out," Woinarski explained, "parties concerned about the well-being of native species managed to convince the Federal Environment Minister that he should draw the line on the extinction of Australia's endemic animals. The greatest opportunity was to campaign against feral cats."

It took roughly twenty years for the notion of controlling feral cats by lethal humane means to become policy in Australia. Humane, in this context, implies "with minimal suffering." Predation by feral cats was listed as a Key Threatening Process under Australia's Federal Endangered Species Protection Act of 1992. This led to the creation of the Threat Abatement Plan [TAP] for Predation by Feral Cats in 1999 by Environment Australia and an updated plan in 2008, created by the Australia Department of the Environment, Water, Heritage and the Arts (as the Australian Department of the Environment was then known). "Although total mainland eradication may be the ideal goal of a cat TAP, it is not feasible with current resources and techniques," the 2008 plan states. "Cat populations must instead be suppressed and managed to mitigate impacts in targeted areas where they pose the greatest threat to biodiversity."[10] Government documents concerning the reduction of cat predation emphasize that reducing cat populations is merited only as a means of protecting and abetting the recovery of native species; it is not killing for killing's sake. The TAP also emphasizes the need for "effectiveness, target specificity, humaneness and integration of control options for feral cats," and highlights the importance of "increasing awareness of all stakeholders of the objectives and actions of the TAP, and of the need to control and manage feral cats." In addition to the TAP and other national initiatives to mitigate the impact of cats on wildlife, legislation has been introduced in many Australian states and territories to restrict the reproductive and predation potential of owned domestic cats. On a local level, many municipalities have introduced legislation that includes the banning of cats as pets in some communities, compulsory spaying or neutering, individual identification, and containment of pet cats.

In July 2015 the Australian government announced plans to cull up to 2 million feral cats by 2020 in a bid to provide a possible reprieve for the more than 100 species of mammals (including the Numbat and species of bilbies, bandicoots, and bettongs) and thirty-plus bird species (including the Spotted Quail-thrush, the Buff-banded Rail, and the Swift and Orange-bellied Parrots) that are threatened with extinction brought on, at least in part, by feral cats. Australian environment minister Greg Hunt declared that "We are drawing a line in the sand today which says, 'On our watch, in our time, no more species extinction.'"[11]

A large part of the culling program will hang on the success of a bait dubbed Curiosity, a poison encased in a skinless sausage that was designed to appeal especially to felines. Adapted from an earlier iteration of cat bait called Eradicat, it combines kangaroo meat, chicken fat, and flavor enhancers with a dose of para-aminopropiophenone (PAPP). PAPP acts by converting hemoglobin in the animal's blood to methemoglobin, which leads to death by inhibited breathing. Death by PAPP ingestion has been likened to falling into a sleep from which one does not wake up. Testing in select regions has shown that nontargeted species—that is, animals that are not feral cats—are not interested in the bait or will spit out the PAPP-containing capsule if they do attempt to eat it. (A method of PAPP delivery through oral grooming rather than bait is also being explored.) The Royal Society for the Prevention of Cruelty to Animals (the British version of the ASPCA) has indicated that cats that ingest PAPP die a humane death. Because feral cats in Australia are very widely dispersed in the most rural areas, where they pose the greatest danger to endemic birds and mammals, management techniques such as trapping or shooting, or constructing large-scale fences to protect wildlife from predation, are not feasible from either an economic or operational perspective. Instead, baits are dropped by plane in critical regions. (It should be noted that baiting is being conducted far from any human population centers, so owned cats who are let outside face little danger of exposure to Curiosity.) Eradication programs are also under way on Australian islands that have feral cat populations, including Christmas Island.

The Australian government has dedicated significant resources toward the eradication efforts—over $100 million in Australian dollars for the program's first four years, starting in 2015. Researchers are also exploring the possibility of introducing viruses into feral cat populations. Some believe this method poses the best chance at widespread eradication.

Why do Australians tolerate the wholesale killing of feral cats? Is there a deep-rooted antagonism in the Aussie character toward felines or, at best, indifference? Not necessarily. "Domesticated cats are much-loved pets for many Australians," said Australia's Threatened Species Commissioner, Gregory Andrews, when asked about Aussie attitudes toward cats. "But there is growing community recognition that free-ranging feral cats have proven devastating to wildlife, especially our small mammals, lizards, frogs and ground-nesting birds. And Australians do value their unique native fauna, and they are important to Australia's cultural identity."[12]

Though a majority of Australians have been accepting of their government's decision to begin feral cat eradication efforts, some outsiders have felt it necessary to voice their opposition. The French actress Brigitte Bardot has said, "This animal genocide is inhumane and ridiculous," arguing that Australia should spay or neuter feral cats rather than kill them. The English pop singer Morrissey also chimed in, calling the Australian government "a committee of sheep farmers who have zero concerns about animal welfare or animal respect."[13] As with some examples we have shown earlier, these very vocal advocates seem to overlook the "genocide" of native animals perpetrated by introduced predators and display little "animal respect" for the Australian Continent's 100-plus threatened species.

❖

There is no doubt that Australians place value upon the native fauna being victimized by feral cats. Meanwhile, on the neighboring island nation of New Zealand, the five species of kiwis still in existence—birds that are revered as a national symbol—face possible extinction, brought on by a host of invasive predators, including stoats, ferrets, domestic dogs, and cats. The kiwis are not alone; a number of other native New Zealand birds face the fate of the Stephens Island Wren (see chapter 1) at the paws of cats and other invasive predators, including the New Zealand Kaka, the Weka, and kokako, mohua, and saddleback species. One New Zealander, an economist turned investor turned philanthropist/social activist named Gareth Morgan, is not happy with the situation—and he has the means and gumption to do something about it (fig. 6.3).

In January 2013 Morgan raised the ire of the cat lovers of New Zealand—a nation said to have more cat owners per capita than any other country—by launching a campaign called Cats to Go. On the surface there is nothing subtle about Morgan's approach. The campaign website begins with the declaration "That little ball of fluff you own is a natural born killer," and goes on to declare, "Every year cats in New Zealand destroy our native wildlife. The fact is that cats have to go if we really care about our environment." Adorned with slightly eerie cartoon cats, the Cats to Go website highlights the damage that free-ranging cats have inflicted upon New Zealand wildlife; warns that all cats that are let outside are hunting, despite their seemingly gentle nature ("The fact is that your furry friend is actually a friendly neighborhood serial killer"); describes a pristine New Zealand without cats; and berates the Society for the Prevention of Cruelty to Animals for condoning trap-neuter-return programs.[14]

The offices of the Morgan Foundation are on the second floor of an older brick building near the waterfront in Wellington, New Zealand, a pleasing city that recalls a smaller version of San Francisco, with steep hills sloping down to Lambton Harbour. There is an open floor plan, and several employees are fixed before large computer screens. Morgan is distinguished by a broad forehead, a sweeping red mustache that extends below his lip line, giving him a slightly dolorous expression, and a mildly mischievous manner. He explained that the genesis for Cats to Go was a project called Our Far South. "We basically took a cross section of New Zealanders from different walks of life, put them on a ship with some scientists, and headed south to the subantarctic islands to expose them to nature," he said, between bites of a breakfast sandwich.[15] Seeing the environment there, the logic went, would raise awareness of the ecological issues facing New Zealand's islands and mainland. One of the new understandings Morgan and his fellow travelers came away with was the impact that introduced predators, particularly rats and mice, had on nesting bird populations on the islands. On some of these islands, steps were being taken to eradicate the rodents that fed on seabird eggs and chicks.

"This got us interested in trying to address some of the invasive predators on some of our own islands," Morgan continued. "We decided to see if we could remove mice from the Antipodes, the only mammalian species on the islands." This campaign, which was dubbed Million Dollar Mouse, sought to raise $1 million (in New Zealand dollars) toward the effort. New Zealanders donated NZ$250,000 toward the project, the World Wildlife Fund kicked in another NZ$100,000, and Morgan matched contributions dollar for dollar. This was enough to launch the project, and the New Zealand Department of Conservation committed to fund the remaining costs. The eradication process, which relies on poisoned bait, is slated to begin in earnest in the spring of 2016. The Million Dollar Mouse project came together pretty easily, so Morgan and his team began looking for a bigger island to tackle. "Stewart is New Zealand's third-largest island, so we began looking at that," Morgan said. "We learned that the invasive predators on the island were possums, rats, and cats. 'Cats?' I asked. 'Are cats a predator anywhere else?' I was told that they were, particularly in townships and cities." And Cats to Go was born.

Morgan likes to joke that the Cats to Go initiative made him the most hated man in New Zealand. It certainly garnered attention to the issue of cat predation, and that was the intention all along. A cat on your lap at home is fine in Morgan's opinion, but he would like to see free-ranging cats removed from the landscape. As in the United States, there is no sanction for free-ranging cats in New Zealand nor any official distinction between owned and unowned cats. Now that the issue of free-ranging cats has entered the public conversation, the Morgan Foundation is pushing for a mechanism that codifies the distinction. Under his pest management strategy, all owned cats would be registered and implanted with a microchip, and a budget would be established to employ a network of animal control officers to collect roaming cats. Captured animals would be checked for a microchip. Owners of cats with chips would be contacted and given several days to retrieve their animal; if they did not, the cat would be destroyed. Animals without a chip would be classified as pests and also be destroyed. The foundation's goal is to lobby regional councils (which determine local policy) to get the program in place to show its viability and then push for national adoption. Management strategies are in place for other invasive predators, Morgan argues. Now it is time for cats.

"People say 'There's no wildlife in the cities, why should we bother?'" Morgan said, gesticulating with coffee cup in hand. "It's not just the concrete that's preventing birdlife from taking hold in the city, it's the bloody cats!" To help illustrate this point, Morgan purchased a set of game cameras and installed them on his property in Wellington. Game cameras are triggered by the motion of a passing animal and are used both by hunters and biologists to monitor the presence of prey and other animals. The first night the cameras recorded nine separate cats crossing his grounds. Other cameras were purchased and placed in yards around Wellington, and the footage showed an astounding number of cat visitations; when extrapolated across the city and for the duration of the year, the number came to 49 million.

In the hills of Wellington, just ten minutes above downtown, rests a unique nature preserve called Zealandia. The nearly one-square-mile sanctuary provides a predator-free home for a host of New Zealand endemic life, including the lizard-like Tuatara, the Hihi (or Stitchbird), the North Island Saddleback, and the Little Spotted Kiwi. Zealandia's success as a preserve is dependent on keeping out predators—including free-ranging cats. To that end, a very expensive fence was designed and built to exclude the roughly thirteen species of nonnative mammals found around the valley. The fence prototype was tested to repel a range of animal capabilities, including jumping, climbing, digging, and the ability to squeeze through tiny spaces; the selected design included three components—a curved top, a wire-mesh wall, and an underground skirt. Completed in 1999, the fence is more than five miles long and completely encloses the Karori Reservoir valley. Its price tag (excluding design) was NZ$2.4 million. Annual costs of maintaining Zealandia eclipse NZ$2 million.

"To me, Zealandia is the most expensive cat-food factory in the world," Morgan said, throwing up his arms. "The birds fly over the fence and *BOOM!*—a cat gets them. I'm constantly asking why we're wasting money on a bird sanctuary if we're not going to take care of the cat problem. I'm trying to get people to think about where their tax dollars are going. People always say to me, 'My cat's good, he/she doesn't kill anything.' If it's wandering around, it's killing things. I had this exchange with the prime minister [John Key], who said his cat, Moonbeam, would never hurt a bird. I said, 'Why not perform an autopsy on Moonbeam. If there aren't any feathers, I'll buy you a new one.'"

If current trends are not reversed, it is expected that at least one of New Zealand's five kiwi species will go extinct within the next fifty years.

🐾

The fates of the kiwis, the Orange-bellied Parrot, and the Piping Plover depend at least in part on whether or not invasive predator species—among them, free-ranging cats—can be suppressed. Beyond the legal, political, and logistical questions such suppression begs, there is a tremendous ethical/philosophical question: Does the fate of a species trump that of the individual? And on another level, does the fate of an individual animal trump that of an ecosystem?

As our world seems to shrink and more and more animals are showing up in places beyond their original range, incidents of culling one animal to save another are becoming increasingly common. In the Pacific Northwest, populations of Pacific salmon and steelhead have been in steep decline, and a number of subspecies are listed as endangered. The Columbia River, which separates Oregon and Washington and is one of the main thoroughfares for these fish as they head back to their natal waters to spawn, has become a battleground. At the river's mouth the U.S. Army Corps of Engineers plans to eliminate up to 26,000 Double-crested Cormorants (*Phalacrocorax auritus*), which prey on juvenile salmonids as they leave the Columbia for the Pacific; 11,000 birds are to be shot, and 15,000 unborn chicks terminated by egg oiling. One hundred and fifty miles upriver, adult Chinook Salmon (*Oncorhynchus tshawytscha*) waiting to negotiate the fish ladders at Bonneville Dam face a different predator, California Sea Lions (*Zalophus californianus*). After cracker shells (loud like firecrackers but not injurious) and other nonlethal deterrents failed to dissuade the sea lions from killing the salmon, wildlife agencies in Washington, Oregon, and Idaho received federal authorization to euthanize problem animals. To date, nearly 100 sea lions have been killed. In each of these cases, native species are being killed in an attempt to save other native species.

For bird lovers, perhaps the most vexing drama involving the killing of one creature in hopes of sustaining another is also unfolding in the Pacific Northwest. In the old-growth forests of northern California, Oregon, and Washington, Barred Owls (*Strix varia*) have been displacing (and in many cases killing) the threatened Northern Spotted Owl (*Strix occidentalis caurina*). Even individuals uninterested in environmental news will likely recall the Spotted Owl controversy of the early 1990s, when logging was curtailed in large swaths of forestland to preserve habitat for the critically endangered bird—a rather diminutive creature, as owls go, and seldom witnessed in the wild. The curtailment led to venomous and at times violent standoffs between loggers, environmentalists, and the officials charged with upholding the cutting bans. Passing through small towns along Highway 101 from Eureka, California, to Forks, Washington, one would come upon restaurant signs reading "Spotted Owl Served Here" and bumper stickers bearing the legend "I Like Spotted Owl . . . *Fried*."

Barred Owls historically made their home in the eastern United States but by 1949 had begun appearing in northern British Columbia. They slowly made their way south, arriving in Washington by the late 1960s, Oregon in the late 1970s and California by the mid-1980s. Barred Owls seek the same habitat as Northern Spotted Owls and, being the larger and more aggressive of the two species—and having a more catholic diet and requiring less territory to survive—have a competitive advantage. Where Barred Owls have colonized, Northern Spotted Owl populations have plummeted.

In an effort to stem the impact of Barred Owls on Spotted Owl habitat, federal officials have green-lighted an experiment to hire marksmen to reduce Barred Owl populations in four limited regions. The marksmen have been authorized to kill as many as 3,600 Barred Owls in the selected regions. One of the designated marksmen, a retired wildlife biologist named Lowell Diller, expressed the profound ambivalence he felt about shooting a bird he would normally revere in a conversation with *National Geographic*. "When I went out to do it the first time, I was shaking, I had to steady myself," he recalled. "I wasn't sure I could actually do it. It was so wrong to be shooting a beautiful raptor like this. It continues to be awkward to this day."[16] As of this writing, Diller has shot approximately 100 Barred Owls. Bob Sallinger, the conservation director of the Audubon Society of Portland, summed up the dilemma of exterminating one owl to save another succinctly: "On the one hand, killing thousands of owls is completely unacceptable. On the other hand, the extinction of the Spotted Owl is completely unacceptable."[17]

A number of ethicists have grappled with the issue of sacrificing one animal for the chance of another to survive. Dr. Bill Lynn (a research scientist in the George Perkins Marsh Institute at Clark University in Worcester, Massachusetts; Senior Fellow for Ethics and Public Policy at the Center for Urban Resilience at Loyola Marymount University in Los Angeles, and former director of the Masters in Animals and Public Policy [MAPP] program at Tufts University near Boston) was retained by the U.S. Fish and Wildlife Service to review the Barred Owl culling initiative. He concluded that much of the responsibility for the decline of the Northern Spotted Owl lay at the hands of humankind, in the form of overly aggressive logging; in more intact old-growth forest, he reasoned, there might be a different competitive dynamic between the two species. Given that humans helped create the problem, Lynn believes it is humans' role to make up for the harm that has been caused to the Northern Spotted Owl, a subspecies nearing the brink of extinction—even if that means killing members of another species. He has referred to the culling as "a sad good."[18]

Another school of thought posits that the conventional approach to conservation—the attempt to maintain or promote conditions that enable native species to survive—risks ignoring the lives and experiences of wildlife. This line of thinking, dubbed "compassionate conservation," draws on research that documents animals' cognitive and emotional states—that crustaceans can learn to avoid pain, for example, and that bees are capable of pessimism.

The leading voice of compassionate conservation is Marc Bekoff, a former professor of ecology and evolutionary biology at the University of Colorado, Boulder, and cofounder, with Jane Goodall, of Ethologists for the Ethical Treatment of Animals. As we learn more about how animals think and feel, Bekoff's logic goes, the less we can ignore any suffering that we inflict upon them. "The life of each and every individual animal is valued," Bekoff has written.[19] And given this guiding principle, Bekoff believes that killing off members of one species to save another is unacceptable.

❧

Back in Texas, the furor over Jim Stevenson's culling of the cat under the bridge at San Luis Pass sparked a debate focused specifically on the ethics of killing free-ranging cats in order to preserve an endangered species. J. Baird Callicott, University Distinguished Research Professor of Philosophy at the University of North Texas and co-editor in chief of the *Encyclopedia of Environmental Ethics and Philosophy*, saw the duality of the situation. Speaking to the *New York Times Magazine*, Dr. Callicott said, "From an animal-welfare perspective, confining cats and shooting the cat, in the Galveston example, is wrong. From an environmental-ethics perspective it's right, because a whole species is at stake. Personally, I think environmental ethics should trump animal-welfare ethics. But just as personally, animal-welfare ethicists think the opposite."[20] Holmes Rolston III, University Distinguished Professor of Philosophy at Colorado State University and author of *A New Environmental Ethics: The Next Millennium for Life on Earth*, shares Dr. Callicott's opinion but is less sanguine. "You're trading a feral cat, an exotic animal that doesn't belong naturally on the landscape, against piping plovers, which evolved as natural fits in that environment. And it trades an endangered species, piping plovers, against cats, which as a species are in no danger whatsoever. Suffering—the pain of the cat versus the pain of the plover eaten by the cat—is irrelevant in this case."[21]

Jim Stevenson spent more than a year awaiting the trial that could have cost him $10,000 and two years of his life. He continued observing birds around Galveston. The trial was held the week of November 12, 2007, at the Galveston County Court House before district judge Frank Carmona. Paige L. Santell, then Galveston County assistant district attorney, said that Stevenson shot the cat in cold blood and that the cat died a slow and painful death, "gurgling on its own blood." She attempted to make the case that John Newland, the bridge attendant, had cared for the cat, providing food, bedding, and toys. He had even given the animal a name, Mama Cat. Stevenson's defense counsel Tad Nelson countered that buying some food and toys for a cat does not make you the owner if you have not taken other steps such as having the animal spayed or neutered or purchasing a collar and tags. Newland's actions suggested a love for cats, not ownership. In the course of arguments, pictures of the crime scene were shared with the jury of eight women and four men. Stevenson's .22-caliber rifle and a magazine of hollow-point bullets were also on display. During a break in proceedings, Stevenson told a reporter that cat fanciers who have condemned him and sent him hateful correspondence, "think birds are nothing but sticks. I did what I had to do."[22]

The trial lasted three days, and jurors deliberated the case for eight and a half hours over two days. In the end, the foreman informed Judge Carmona that they were deadlocked, and on November 16 the judge declared a mistrial. Shortly after, Galveston County District Attorney Kurk Sistrunk announced that Stevenson would not be retried. After the decision was announced, Stevenson was hopeful. "I think it's a real positive step from the DA's office because it shows they are making some progress in bringing bird and cat lovers together and will save a lot of money," he told reporters.[23]

Whether bird and cat lovers were brought together by the decision—or lack of a decision—is debatable. A posting on *Cat Defender*—a popular blog among free-ranging cat advocates—did not exactly strike a conciliatory tone:

> Bird lovers all over the world are still whooping it up following serial cat killer James M. Stevenson's great victory last Friday afternoon in a Galveston courtroom. Even the usually dour cat-hating monster found it difficult to contain his elation as he strutted out of court all the while laughing up his blood-soaked sleeve at the travesty of the American judicial system.
>
> After deliberating a scant eight and one-half hours over two days, a panel comprised of eight women and four men told Judge Frank T. Carmona that they were hopelessly deadlocked and he then declared a mistrial. Although the sadistic killer needed only one bird lover or ailurophobe to vote in his favor in order to produce a hung jury, in this instance he got four.[24]

The furor stirred up by *State of Texas* v. *Stevenson* had at least one result. On September 1, 2007, the Texas legislature altered the statute concerning animal cruelty to eliminate the ownership requirement for prohibiting the killing of a domestic animal. "Feral" cats were specifically protected in the new statute.

Had Jim Stevenson shot Mama Cat on September 2, 2007, he would almost certainly have been found guilty. Much to Stevenson's chagrin, in March 2015 the Galveston City Council voted six to one to approve a proposed trap-neuter-return ordinance that would no doubt legitimize the presence of many more Mama Cats on the dunes along Galveston Bay.

CHAPTER SEVEN

Trap-Neuter-Return: A Palatable Solution That Is No Solution At All

It's so much easier to suggest solutions when you don't know too much about the problem.

—Malcolm Forbes

The lobby of the Oregon Humane Society (OHS) facility in Portland is a pleasing space, with high ceilings and big windows that let summer sunlight pour in. Near the front of the lobby is the Kitten Colony, a little glassed-in room where kittens play. One July day a gray kitten tussled with a calico, as a black kitten with a white neck watched from a windowsill within the enclosure. Another white kitten batted a feather toy about. An eighteen-month-old girl sat spellbound outside the window watching the activity, occasionally tapping the window and calling, "Hi, kitty!" All the fun can be streamed to your desktop computer via OHS's Kitty Cam. The kittens are eligible for adoption, and given the attention they garner from visitors and their general level of cuteness, one suspects their odds of finding a home are quite good.

Fifty yards farther into the OHS building, some eighty-three cats rested in stainless-steel cages and on surgical tables in various stages of sedation. If all went as planned, there would be fewer kittens in greater Portland in the near future, as all of these animals were waiting to be or recovering from being spayed or neutered, procedures that will terminate their ability to reproduce.

There are not many facets of feline management on which bird and cat advocates agree, but one point universally acknowledged is that there are simply too many free-ranging cats. Groups ranging from the American Bird Conservancy to the Humane Society of the United States agree that the most effective and humane way of controlling cat populations is to spay and neuter as many cats as possible. As invasive procedures go, these are fast (four to six minutes to spay a female, less than a minute to neuter a male), inexpensive (fees are $42.50 for spayings at the Oregon Humane Society, $32.50 for neuterings), and readily accessible from providers ranging from nonprofit humane societies to city- or county-run animal shelters to private veterinary practices. Sliding-scale fees (down to free) make the procedures affordable for people from any socioeconomic background.

Just about everyone agrees it is important for cats to be spayed and neutered. Sterilization among owned pets is very high; the Humane Society of the United States estimates that 91 percent of owned cats have been fixed. The percentage of free-ranging cats that have been fixed is unknown, though much lower—as low as 2 percent according to some estimates. Where opinions diverge is on what to do with free-ranging and unowned cats once they have been sterilized. Most outside-cat advocacy groups—and a surprising number of municipalities and mainstream animal welfare organizations—have embraced an approach called trap-neuter-return (TNR). TNR works as its name implies. Volunteers trap free-ranging cats, generally at sites where the cats are fed by caregivers (fig. 7.1). The cats are then taken to a clinic, where veterinarians remove animals' testes or ovaries, rendering them unable to reproduce. Once revived, the animals are returned to the site where they were trapped to live out their days.

Sarah Smith, a sixty-something volunteer, has devoted an enormous amount of time and resources to capturing outside cats, conveying them to clinics for sterilization, and returning them to the farm areas and apartment complexes where they were found in the first place. ("Sarah Smith" is a pseudonym, used to protect the privacy of the person profiled.) She has been practicing TNR for over ten years, most recently in Oregon's Willamette Valley, to which she relocated from the East Coast, with seven rescued cats—and her husband. She had not planned to get involved with free-ranging cats in Oregon but found she could not stand to see their suffering; as an animal lover, she felt she had to do something.

Cats were an acquired taste for Smith. She grew up in the Midwest, and all her relatives had dogs. When she first moved to the East Coast, Smith had a roommate who had two cats, but Smith did not care for them. Eventually she married, and when she and her husband moved to the suburbs, they got Petey, a Yorkshire terrier. Petey's presence in the household contributed to Smith's "cat-tharsis" and her commitment to trapping. During a big thunderstorm one evening, Petey kept going to the door and whining. Finally, Smith went to the door and could make out a *meow* from the bushes in front. She felt around and pulled out a little black-and-white kitten, which she took in. Conversations with neighbors and a little investigative work revealed that a free-ranging cat had given birth to a litter of kittens in a nearby shed; "Lucky" had probably been dropped in Smith's bushes as the mother was moving the kittens to a different spot. Smith called a local cat advocacy group and borrowed a trap, hoping to catch and spay the mother cat.

"That first attempt at cat trapping, I didn't have any luck," she recalled. "A few months later, I saw the momma cat running down the sidewalk. A neighbor and I followed, and we found yet another litter of kittens in a nearby pool pump house. Catching this female turned out to be quite difficult. So I decided to gather all of her kittens in a cat carrier first and then place a trap outside the pump house along with the cat carrier containing the kittens. Finally, after several days of bottle-feeding kittens and unsuccessful trapping, I looked out the window and saw that the trap's door had gone down. It was the momma cat. I took her in to a local clinic and got her fixed and then released her and her kittens back behind the pump house."[1]

The exact time and place that TNR was first implemented as a potential population control measure is unclear, though by the

1970s it was being selectively practiced in England and Denmark. The earliest formal TNR efforts in the United States date back to the early 1990s, though the practice was discussed in the feline press (but not by name) as early as 1984. Proponents tout it as a way for individual populations of free-ranging cats to slowly disappear through a more "natural" death process rather than through the six- to twelve-second respiratory arrest experienced by an animal when a veterinarian or humane society employee injects a dose of sodium pentobarbital during the euthanasia process. TNR advocates believe that the practice also improves the lives of cats, relieving them of the intermittent stresses of mating and pregnancy, and improves their physical health, as cats are generally given vaccinations against distemper, herpes virus, feline calicivirus, rabies, and sometimes feline leukemia virus. Proponents also argue that TNR provides the humans who happen to live in proximity to colonies some relief from troublesome cat mating behaviors like roaming, caterwauling or yowling, urine marking, and fighting.

A case could be made that TNR makes life a bit more bearable for free-ranging cats, though it still leaves them to face all the challenges of living in an outdoor environment they are ill-equipped to face. But returning neutered cats to their colonies also returns them to preying on any animals they can catch and kill, an instinctive behavior they cannot resist. From a conservation perspective, this is unacceptable. And even though TNR cats may receive some vaccinations, very few receive boosters. The absence of boosters leaves the animals susceptible to diseases, which can be passed on to other cats, wild animals, and humans. From a wildlife and public-health perspective, this too is unacceptable.

There is also the uncomfortable truth that TNR has repeatedly been shown to fail to reduce free-ranging cat populations (fig. 7.2).

❧

TNR and its no-kill tenets have their roots in the animal liberation movement, a crusade that began to take shape in the pubs and hallowed halls of Oxford University in the late 1960s–early 1970s. Eminent philosophers such as Peter Singer, Richard Ryder, and

Richard Hare began to argue that animals—that is, nonhumans—should be afforded the same rights as people. The reasoning was that every nonhuman animal had inalienable rights as a living thing because each animal could suffer much as humans do. To fail to account for the rights of animals, they argued, would be a form of "species-ism," or animal discrimination. Suddenly, wearing fur was not a statement of wealth but an act of cruelty. So were participating in a rodeo, eating a steak, all forms of animal research, and holding up your trousers with a leather belt.

Thanks to several influential books, notably *Animals, Men and Morals*, and *Animal Liberation*, the movement gained momentum in Europe, the United States, and Canada. One of the most visible expressions of the movement's legacy is found in People for the Ethical Treatment of Animals, or PETA, which was founded in 1980. As the organization notes in the history section of its website:

> Before PETA existed, there were two important things that you could do if you wanted to help animals. You could volunteer at a local animal shelter, or you could donate money to a humane society. While many of these organizations did useful work to bring comfort to animals who are used by humans, they didn't question why we kill animals for their flesh or their skins or why we use them for tests of new product ingredients or for our entertainment.[2]

PETA, of course, has gone on to champion a broad range of animal rights causes; for example, the group fights for bringing to a halt the use of animals in product testing, better living conditions for livestock slated for slaughter and consumption, and a moratorium on the use of fur in fashion products. (PETA's "I'd Rather Go Naked Than Wear Fur" advertising campaigns and red paint–drenched fashion shows set a high bar for animal welfare media hijinks; see fig. 7.3.) PETA, it is worth noting, does not support TNR:

> Sadly, our experience with trap, spay-and-neuter, and release programs and "managed" feral cat colonies has led us to question whether or not these programs are truly in the

cats' best interests. We receive countless reports of incidents in which cats—"managed" or not—suffer and die horrible deaths because they must fend for themselves outdoors. Having witnessed firsthand the gruesome things that can happen to feral cats, we cannot in good conscience advocate trapping and releasing as a humane way to deal with overpopulation.

Advocates argue that feral cats are just as deserving as other felines and that it is our responsibility to alleviate their suffering and assure their safety. We absolutely agree. It is precisely because we would never encourage anyone to let their own cats outdoors to roam that we do not encourage the same for feral cats. In fact, the act of releasing a feral cat is, in the eyes of the law, abandonment and is illegal in many areas.

We believe that although altering feral cats prevents the suffering of future generations, it does little to improve the quality of life for the cats who are left outdoors and that allowing feral cats to continue their daily struggle for survival in a hostile environment is not usually a humane option.[3]

PETA stops short of advocating that all free-ranging cats should be euthanized, but the group's position underscores the point that such animals lead grim lives and that returning animals to such a predicament borders on cruelty.

Back at the Oregon Humane Society two vets—Margaret Wixson and Wendi Rekers—strode purposely about the Coit Operating Suite of the 4,000-square-foot Holman Medical Center, trailed by several veterinary techs. Everyone was attired in scrubs. The center averages more than 12,000 surgeries annually, and boasts a near-zero euthanasia rate for pets who arrive at the shelter in need of care, even though most owners cannot afford to cover the cost of procedures. Every pet that OHS takes into its shelter is spayed or neutered, and the center also sterilizes thousands of other owned pets for a nominal fee, or free of charge, as part of the organization's Spay & Save program for low-income families. Holman Medical Center is a teaching facility, through which veterinary students from Oregon State University College of Veterinary Medicine cycle every three weeks. "Dr. Wixson interned here last year, and decided to stay on as an employee," Ron Orchard, the shelter's medicine manager, explained. "We're lucky to have her."[4]

From an observation window outside the operating suite, we could watch two cats—each on its own table—in the process of being spayed. Spaying is the desexing procedure for females (fig. 7.4). The cats lay on their backs, legs secured by blue ties attached to the table to hold the animals steady during the procedure. Before being delivered to the operating suite, the cats are sedated—a cocktail of morphine and a tranquilizer followed by an injection of propofol. Then they are intubated; once on the table, they receive oxygen mixed with a small percentage of anesthetic gas to keep them sleepy through the procedure. Each cat's paw is fitted with a monitor to measure heart rate and oxygen saturation. Once the cats are secured, they receive a sterile scrub, three swabs each of alcohol and chlorhexidine, and a squirt of iodine for good measure. At this point, one of the techs opens a sterile pack, which includes a scalpel and other surgical tools. A suture pack is opened and placed on the table, and a towel is placed over the animal's upper torso to keep her warm. A drape is then placed over the cat's body to create a larger sterile field; there is a small hole in the drape for the area over the abdomen.

Dr. Wixson approached the table on the right and picked up the scalpel. Reaching through the hole in the drape, she made a tiny incision in the skin, not more than a few millimeters; the extremely sharp knife cut through easily. She then cut through the animal's body wall and removed a bit of fat. Picking up the spay hook, a longer instrument with a small hook on the end, Dr. Wixson swept it through the cat's abdomen, trying to loop the hook around the uterus. "This is the toughest part of the process," Orchard explained. "The vet sometimes has to hunt around a bit until they find the uterus." Dr. Wixson quickly located the uterus, a Y-shaped section of thin pink tissue, and slowly worked both segments of the Y out onto the drape. The cat's ovaries are attached at the end. She ligated the bottom of the uterine body to minimize bleeding. Another cut below the forks of the Y, and the uterus and ovaries were removed. No blood was shed. It took only a few stitches to close the body wall. A small tattoo was applied—a tiny green line—to indicate that the animal had been spayed. After a small bit of glue was applied to seal the skin, a vet tech carried the still-unconscious cat from the operating room to a table in the immaculately clean main room of the center and laid it down upon a warming pad. Another tech was there to attend to the cat as it returned to consciousness. Once the cat moved a bit, the breathing tube was removed. The tech gently rubbed the cat and spoke softly to ease it back into consciousness. "The wake up period can be the most critical part of the process," Orchard said, as the kitten in question slowly opened its eyes and tried to struggle to its feet. It was then brought back to a cage with its identifying card—white, to show it is a shelter cat, now ready for adoption.

The process of neutering—desexing of male cats—is less invasive and thus much faster. Male cats are sedated (but not intubated), and then a tech shaves their scrotums and sterilizes the region with one scrub each of alcohol and chlorhexidine. Then the cat is brought to a table with a bright light positioned above. The vet picks up a small scalpel no longer than a pinkie finger and makes two tiny incisions. The cat's testes are pulled out, the spermatic cords are tied off, and the testicles are severed. Then the cords are tucked back inside the animal, and it is sutured up. The procedure is done in less than thirty seconds.

The Oregon Humane Society does not practice TNR. "We don't return sterilized animals to the outdoors," Orchard explained. "But we respect the roles of other animal advocacy groups. TNR is the Feral Cat Coalition's domain." (This organization focuses on training TNR practitioners and performing spay/neuter procedures for free-ranging cats.) When asked about the appeal of his job providing medical services to animals—some of which have been neglected or mistreated—Orchard did not pause a moment. "It's a super feeling to be able to help. But it's unfortunate that we can't help them all."

Sarah Smith certainly feels that she is helping to make a difference in the lives of the cats she traps. She conducts most of her trapping on Thursdays, as the clinic where she generally takes the cats—Willamette Humane Society—provides spaying, neutering, and vaccination services on Fridays. The service is provided for a donation of $43, which is covered by Salem Friends of Felines, a nonprofit cat advocacy group that underwrites much of its expenses through a sizable grant from PetSmart Charities, the charitable arm of the pet store chain. Smith learns about colonies of cats that might be in need of TNR services through Friends of Felines; individuals in the community contact Friends of Felines for help, and the group in turn reaches out to volunteers like Smith.

One Thursday afternoon in April 2013, Smith arrived at a two-story low-income apartment complex behind a Red Lion hotel and adjacent to Interstate 5 in Salem, Oregon. Approaching one of the units, she spied several plates out with remnants of cat food along the edges. She sighed. "We have a snowball's chance in Hell of trapping any cats tonight," she said.[5]

A resident of the complex—the woman who had called Friends of Felines seeking TNR assistance—approached her. She is in her mid-fifties, a home health attendant, and was agitated. "I took over the feeding this week," she said. "Regular amounts on Monday and Tuesday, light yesterday. But someone else fed them today. I wanted to scream when I saw the feed bowl." After a moment she added, in a lower voice, "We have some people with mental health issues here."

Smith carried five live traps out of the side of her minivan and fanned them out along the back edge of another shabby apartment unit, where several open casements gave way to a crawl space below. The traps are manufactured by Tomahawk Live Trap, which was founded in Tomahawk, Wisconsin. Each is thirty inches long by ten inches wide by twelve inches high and built from one-and-a-quarter-inch mesh in the United States. There is an opening in the front for the cat to enter and a sliding door in the back where users can slide food in. The trap's trip pan is roughly a third of the way in; when the cat steps on the pan, the door closes, and the animal is trapped. Tomahawk also manufactures traps for armadillos, badgers, bats, beavers, birds, Bobcats, chickens, chipmunks, Coyotes, crayfish, dogs, foxes, gophers, groundhogs, ground squirrels, jackrabbits, mice, moles, muskrats, opossums, pigeons, prairie dogs, rabbits, Raccoons, rats, reptiles, roosters, shrews, skunks, snakes, squirrels, turtles, voles, and woodchucks. Smith lined the bottom of the traps with newspaper and then opened a few cans of cat food. She spooned the food onto some plastic plates and set them in the traps.

"You should tell the people that put the food out that they're not being fair to the kittens by letting them breed," she said, turning to face the woman. "Next time I'm out, I may as well have a gun. Animal control will be next. The management company will hire an exterminator to kill these cats. Someone will complain about the smell. And it will only get worse when they mate again."

"These cats aren't into mating," the woman replied.

"They're all into mating," Smith said curtly.

This venue is not atypical of the places where feral cats tend to congregate in the Salem area. Smith has certain conditions for the cat colonies and caretakers she will work with: the caretakers must be committed to continue feeding and otherwise providing for the cats after they are neutered or spayed and returned, and she will not work with people who just want the cats removed. Many of the people with whom Smith works lack disposable incomes and will forgo small comforts for themselves, such as dinner out, to provide for the neighborhood cats. In a study in which a number of TNR participants were asked about their motivations, "love of cats," "opportunity to nurture," and "increased self-esteem" were among the top reasons.[6] Smith noted that she encounters very few wealthy people who actively support cat colonies; people who have some means tend to write a check instead.

After setting up the traps, Smith returned to her van—roughly twenty feet away from the nearest trap—rolled up the windows, and tuned into the local jazz station at a low volume. And the wait began. Trapping cats is a lot like fishing or hunting; lots of preparation and waiting, with the occasional reward of a capture. When asked about cats' proclivity for hunting, she paused. "It's easier to make cats out as the bogeyman than to look at larger problems we have. We're degrading forests, polluting the water. If cats are the only problem with birds, why are butterflies and bats dying out? No one wants to deal with the bigger problem. Cats don't have a chemical industry behind them. Aren't there bigger issues to worry about?"

Ten minutes later, a black-and-white cat stuck its head out of one of the casements. It walked around one of the traps, then sat down and yawned. Up again, the cat slowly inspected another trap, sticking its head inside the trap a bit, then stepped out before inspecting another trap, sniffing its contents from the side. "'There's food in this one, I'll sniff it,'" Smith said, channeling the cat's thought process. "'There's food in that one, too, I guess I'll sniff that.' I didn't realize that cats like to sniff as much as dogs. They won't walk in a trap unless they're really hungry." The cat returned to the crawl space for a minute, then crawled out again. It seemed to be looking at the van with its green eyes, cocking its ears as if listening to the music, though the windows were closed.

Smith keeps detailed records of the cats she traps and has had spayed or neutered. In 2012 she trapped 240 cats. Her best haul on a single evening was fourteen; the average is four or five. Smith estimates that there are perhaps ten other people around Salem who do TNR. It is not clear how many managed cat colonies there are in Oregon or even in the mid-Willamette Valley.

In its early stages of practice, TNR was something of a fringe movement, embraced and espoused by volunteer-driven outdoor-cat advocacy organizations like Alley Cat Allies and the Feral Cat Coalition. But now the activities of Sarah Smith and other like-minded caregivers have entered the mainstream, receiving enthusiastic endorsements from animal welfare organizations such as the Humane Society of the United States and the American Society for the Prevention of Cruelty to Animals (ASPCA).

Many government entities have gotten into the TNR game, endorsing such programs and in some cases even underwriting them. Take the city of Houston, for example, which makes the case for TNR on its website:

> For a long time, "catch and kill" was a widely accepted method of managing community cat colonies. The cats were trapped and removed from their established colony to be euthanized.

While this method causes an instant decrease in the overall colony numbers, *it is not effective over time*. Colonies subject to "catch and kill" typically end up increasing in number back to their original size as a result of what is known as the *vacuum effect* [emphasis added; more on this subject below].

Once the community cats within a colony are spayed and neutered, not only will the population size gradually decrease, but the cats will also be healthier and coexist more peacefully within a neighborhood. Female cats, prevented from having any more litters, will be healthier. Male cats will gradually lose the urge to roam and fight, and will be less prone to injury. Behaviors associated with unaltered cats, such as yowling and marking territory with urine, will disappear.[7]

Anticipating objections from members of the birding/ecology community, the city addresses the following question: Does putting community cats back into the community increase the risk for birds and wildlife to be harmed?

It has been argued that cats should be collected from the community, impounded and euthanized in shelters to protect wildlife and public health. However, euthanizing or removing all community cats from an area may lead to an increased population of other non-native species with an even more detrimental effect. There are many more cats in the community currently than BARC [the city of Houston's animal shelter and adoption facility] can take in over a short period of time. The TNR program will decrease the number of cats that could potentially harm birds and wildlife over time.[8]

San Francisco has also been promoting TNR, pointing interested participants to the San Francisco Society for the Prevention of Cruelty to Animals, which spearheads efforts in the city. The SPCA has devoted considerable resources to promoting its TNR program under the rubric Community Cares. Laura Gretch, a vivacious and much-tattooed forty-something, managed the program from 2009 to 2014. "We began with the idea that fewer cats on the street equals less of an impact on bird and other animal populations," she said from her office on the less fashionable side of Potrero Hill. "Whether people agree with TNR as a solution or not, you can't deny that it does reduce the number of cats on the streets." (Actually you can—more on that later.) Gretch was quick to admit that feral cat populations are hard to pin down. "If you sit in one place and try to count, you'll inevitably start counting twice. However, we can count the number of cats that we've spayed and neutered, and that's one cat that's not reproducing. We've had community cat caregivers report that they have seven cats that they feed. The cats were captured, spayed/neutered and returned. We came back and counted five years later, and there are only three cats. This would suggest the program is working, though it's hard to know."[9]

Under Gretch's guidance, the San Francisco SPCA took its TNR campaign to the streets. "I'd call the spay/neuter program aggressive," she said. "We wake up and want to see how many cats we can fix." Gretch and her communications team wanted to steer away from any hints of preachiness. "There's a lot of didacticism in animal welfare circles, and that gets ignored or turns off the average Joe who's not interested in community cats or doesn't even know what they are." Instead of telling the community what they should think or feel about these cats—or that they should think anything—the SPCA developed a simple campaign titled Do You See Cats? The hope was to open up a conversation and, at the same time, ask people to help in a simple way, without confusing them. The message seemed to resonate. "Whether you like cats or not, you might call to say you saw some," Gretch explained. "We thought this would play well in a city like San Francisco, where many people want to get involved. People tell us where they see the cats; we take it from there." The campaign ran advertisements on buses and bus shelters and sent out direct mail pieces, all in Mandarin and Spanish as well as English. The copy below the campaign slogan informed readers that the SPCA would spay or neuter cats free of charge and listed a phone number.

"While the campaign was running, we doubled the number of cats we fix, from just under 1,000 a year to over 2,000," Gretch said. She is sincerely committed to her TNR mission. For several years she auctioned off a section of her skin to be tattooed with a TNR-centric message. The winning bidder got to choose a tattoo to join the many that already festooned Gretch's body. An inking from the 2012 auction reads "Spaneuter" (fig. 7.5).

San Francisco SPCA's ad hoc data points and "statistics"—which are cited by thousands of municipalities, humane societies, SPCA branches, and cat advocacy groups across America—purport that TNR reduces free-ranging cat populations (and by association their impact on wildlife). The truth is that trap-neuter-return makes people—not cats and certainly not wildlife—feel better. It gives individuals and government entities the affirming sense that they are taking action, while allowing them to escape facing any of the hard decisions that reasoned action demands. There is a dearth of good science supporting the claim that TNR works to reduce cat populations. And there is considerable evidence that suggests it does not.

One example that is frequently trotted out to demonstrate the efficacy of TNR for cat colony decline comes from a ten-year (1991–2001) study by Julie Levy and colleagues on the campus of the University of Central Florida in Orlando. The study combined two different approaches—TNR and TNA (trap-neuter-adopt)—which blurred the researchers' ability to carefully assess the efficacy of TNR alone. UCF has a large, wooded campus, and as happens in many places with transient populations (college campuses, military bases, field stations), cats are often left behind as people come and go. Before the study began, the university was dealing with a growing unowned cat population. In 1991 volunteers with the group Friends of Campus Cats began a TNR program to reduce feline populations. Feeding stations were established at eleven different areas around campus to concentrate cats. This created eleven different colonies. Cats were then captured and transported to veterinarians for neutering and vaccination. Any cats testing positive for feline leukemia virus or feline immunodeficiency virus were euthanized. The tip of the ear was clipped in cats that were returned to the outdoor environment to allow easy field identification. The Levy et al. report includes only one brief mention of a census of the colony, in 1996, and it includes no details of how the count was done. Counting free-ranging cats is notoriously difficult; for a study aimed at examining the impact of TNR on the number of outdoor cats, the absence of details on census procedures suggests a first serious flaw.

Levy et al. reported a significant decline in the number of cats, from the initial census of 155, over the course of the study and attributed this to TNR and to the adoption program. Here is what happened to the cats: Seventy-three of the 155 (47 percent) cats were adopted. This is a wonderful result, but it is not a test of TNR. That brings the cat population for the TNR study to eighty-two. Seventeen of these animals were euthanized for various reasons, most at the beginning, but a few others over the course of the study. That leaves sixty-five outdoor unowned cats. Of these, ten were found dead; six were hit by cars and four died from unknown causes. Nine cats apparently were known to have left and gone into the woods, although it is not clear what that means. Twenty-three cats (42 percent) were "lost," so their fates were also unknown. (They could have dispersed into the woods, they may have died, and their carcasses were never found, or perhaps they were adopted—although this is doubtful.) This left a final population of twenty-three cats. But again, Levy and colleagues do not explain how or when the cats were counted, so it is not clear how reliable any of these figures actually are. What the Levy study shows is that cat colonies will reduce in size if 60 percent of the cats are adopted or euthanized, and if another 21 percent emigrate from the colonies. It says nothing about the efficacy of TNR.

A more careful study of TNR comes from a dissertation completed in 2006 at North Carolina State University in Raleigh. Felicia Nutter, a board-certified veterinarian, who had wildlife experience in Gombe National Park in Tanzania researching baboons and chimpanzees, began a study in Randolph County, North Carolina, in 1998 to determine the survival of outdoor and unowned cats. Her study design involved placing cats into three random treatment groups: (1) unneutered, (2) castrated (testicles removed, resulting in loss of production of sex hormones), and (3) vasectomized (vasa deferentia severed and tied, production of hormones such as testosterone retained). Nutter's dissertation research was unique, differing from most TNR colony studies in that she captured 98 percent of the cats in her study colonies (a rate rarely achieved in practice and perhaps possible in this case only because it was the prime focus of her dissertation and occupied much of her time), and she also added the interesting twist of treating the cats with two different types of sterilization. What she found after four to seven years was that most unneutered colonies increased in population size, whereas colonies receiving both types of sterilization, and experiencing low rates of immigration of new members, significantly declined. As she had predicted, colonies with vasectomized males declined more rapidly than colonies with castrated males. Presumably cats still able to secrete reproductive hormones are prone to fighting. What killed the cats in the two sterilization groups? Nutter assigned a cat the status "dead" when either it simply disappeared (not distinguishable from emigration) or more commonly was killed from trauma (struck by cars, mauled by dogs). Such an ending contradicts the suggestion that TNR is beneficial from an animal welfare perspective. From Nutter's study one can conclude that TNR *can* work to reduce colony populations but *only if* nearly 100 percent sterilization rates are achieved and there is little or no immigration into the colony.

Only one truly rigorous analysis has examined the impact of TNR in reducing free-ranging cat populations from a long-term quantitative and population modeling perspective. The effort was headed by Patrick Foley, a theoretical population biologist in the Department of Biological Sciences at the California State University at Sacramento. The stated objective of the study was to use data from two large and long-term TNR programs to mathematically assess whether TNR was successful and to determine the rate of neutering necessary to cause colony decline. The first program involved TNR data collected by the Feral Cat Coalition in San Diego County, California, where cats were studied from 1992 to 2003. The second program analyzed TNR data collected by a group called Operation Catnip, Inc., located in Alachua County, Florida, where cats were studied from 1998 to 2004. Over the course of the San Diego study, a total of 14,452 unowned colony cats were submitted for neutering to veterinary clinics under the aegis of TNR programs. Of these only about 5 percent had already been neutered. In Alachua County a total of 11,822 unowned cats were submitted for neutering as part of TNR programs. Of these only 2 percent had already been neutered. In neither the California nor the Florida colony had TNR efforts achieved colony decline. Both colonies experienced ongoing population growth and never achieved the high rates of neutering needed to achieve population decline. Foley et al. estimated the necessary rates to be 71 percent in California and 94 percent in Florida. Such levels of neutering, the authors conclude, are unrealistic to attain. Thus, TNR will not lead to colony extinction. This study, unlike any other to date, was long-term, had sufficient sample sizes, and covered a large spatial area rather than a single site, all factors contributing to the credibility of these results.

There are two primary reasons TNR typically fails to reduce free-ranging cat populations through attrition. The first is that caretakers fail to trap and neuter enough cats. Foley's mathematical models showed that 71 percent to 94 percent of a population must be desexed for the population to decline, and this is assuming that no new cats have joined the colony. This level of neutering has not been consistently documented and is extremely difficult to achieve in the field. The Nutter study described above is a rare exception. The second reason is that most colonies are constantly receiving new animals. Most cat advocacy groups argue that colony populations remain stable and resist immigration from surrounding regions, though the scientific literature—and even anecdotal reports from colony caretakers—refutes this argument. Research has shown that cats regularly move between established colonies, and that territories are not defended from interlopers so long as a regular food source is available.

Opponents of lethal removal programs—that is, trap and euthanize—argue that such efforts do not work to reduce colony size because of the "vacuum effect" that draws new cats to colonies from which members have been removed. Not unique to cats, the vacuum effect is based upon the idea that territorial animals residing in high-quality sites (i.e., those with resources like food and shelter) will exclude certain—presumably weaker and otherwise inferior—individuals from that area. This sets up a "winners and losers" situation, in which the excluded animals are forced either to be transient or to occupy nearby territories without adequate resources. TNR advocates claim that, thanks to the vacuum effect, there is a never-ending supply of cats to be "sucked in" to the preferred areas when cats are removed. Given this assumption, removing cats is a pointless exercise if the objective is to reduce the overall population of the cat colony in question.

There are several problems with this logic. First, domestic cats do not commonly exhibit territorial behavior. Second, there is no such thing as an infinite supply of cats (or any other animal, for that matter). If cats continue to appear, it is likely due to the presence of a feeding station—the very resource, of course, that led to the higher than expected population of cats in the first place. Third, and perhaps most important, is the idea at the core of TNR—that colonies will decline due to natural mortality. Wouldn't the vacuum effect that (purportedly) draws new cats into the colony when members are trapped and removed also be at work when a member of the TNR colony dies?

Recent field research from Israel on cat colony population fluctuations suggests that the vacuum effect may indeed exist, yet not the way TNR advocates would like to think. In 2011 Idit Gunther and colleagues studied cats in four feeding groups, two groups receiving TNR and two groups of cats left untreated. The authors then monitored rates of immigration, emigration, and kitten survival between neutered and unneutered groups on a weekly basis over an entire year. They found that the number of adult cats in the two neutered groups increased significantly over the yearlong study period thanks to higher immigration and lower emigration rates than were recorded in the two unneutered groups, where the population decreased. In this case, the vacuum effect actually *increased* the number of animals in existing *TNR* colonies. This, of course, will not always be the result. What one can conclude from this study and other research on the vacuum effect is that replacement will sometimes occur after animals have been removed and sometimes it will not. The outcome is dependent upon how many cats are in the surrounding area, the degree to which those cats are reproducing, and the number that are behaviorally dominant, as well as how many cats are being abandoned.

The veterinary profession has also grappled with the conundrum posed by TNR. David Jessup, a past recipient of the UC Davis School of Veterinary Medicine Alumni Achievement Award, has argued that the American Veterinary Medical Association cannot condone TNR given the organization's stated goal of having a positive impact on the health and well-being of all living creatures. After all, TNR is barely advantageous to one species—cats—and clearly disadvantageous to many dozens, and perhaps hundreds, of other species. Jessup wonders how veterinarians advocating for TNR can justify being party to what amounts to abandonment, an illegal act of animal cruelty. He also worries about the message a pro-TNR stance sends to millions of conservationists and the veterinarians who provide care for birds, native species, and their ecosystems. Another vet, Paul Barrows, chief of the U.S. Army Veterinary Corps, retired, also has advocated for removal without return as the most responsible course of action regarding free-ranging cats. "We must seek to make it politically incorrect and socially unacceptable to engage in biological littering resulting from irresponsible cat ownership and promotion of TNR programs," he reasons.[10]

In an article published in *Conservation Biology*, Travis Longcore, Catherine Rich, and Lauren Sullivan sought to set the record straight on the inefficacy of TNR for reducing free-ranging cat populations by performing a careful analysis of the available scientific (and nonscientific) literature. One of the key conclusions they reached is that TNR is usually framed as an animal welfare problem rather than an environmental problem. A "successful program," given this context, is defined by the welfare of the cats, not the elimination of the free-ranging cats from the environment. (Welfare, it is implied, means staying alive.) They cite one study that concludes that "the effectiveness of the program was demonstrated by the low turnover and improved health of the colony over the 3-year period," though the colony size decreased only from forty to thirty-six. They found that a county in Florida had implemented TNR "to decrease the number of healthy cats euthanized, decrease the costs to the county, and decrease complaints."[11] In such contexts the input of scientists and conservationists is generally not even brought to the table when the subject of TNR is broached, and claims made by free-ranging cat advocates often go unchallenged—and slowly take on the patina of truth.

The scientific community has been wringing its collective hands over the free-ranging cat community's successful whitewashing of the shortcomings of TNR for some years. Given the pro-TNR stance that more and more governing bodies have assumed—and the lack of understanding among the general citizenry about the impact cats are having on wildlife and public health—there is little question that cat advocates are winning the war in the court of public opinion. It is not as if conservation and ecology experts have been unaware of TNR's shortcomings or have not taken positions against the practice. Many influential entities, including the American Association of Wildlife Veterinarians, the American Ornithologists' Union, the American Society of Mammalogists, the National Wildlife Federation, and the American Bird Conservancy (ABC), have come out in opposition to free-ranging cat colonies and TNR. Yet only ABC has made significant investments in educating the public, with its Cats Indoors Campaign. (It should be noted that the budgets for ABC's advocacy efforts pale next to those of such organizations as Alley Cat Allies, Best Friends, and PetSmart Charities.)

A curious omission from the list above is the National Audubon Society, which has a stated mission to "conserve and restore natural ecosystems, focusing on birds, other wildlife, and their habitats for the benefit of humanity and the earth's biological diversity." National Audubon has not taken a firm position on TNR, beyond a board of directors' resolution approved in 1997, which stated, in part:

> National Audubon Society will convey such science-based conclusions [on the impact of feral and free-ranging cats] to its chapters so that they, if they so wish, will be in a position to advocate that local and state wildlife agencies, public health organizations and legislative bodies restrict and regulate the

maintenance and movement of feral and free-ranging domestic cats out-of-doors and to support programs to vaccinate cats and to neuter or spay cats.[12]

The United Kingdom's Royal Society for the Protection of Birds not only takes no position on TNR, it goes so far as to dismiss the impact of cats on birds as largely irrelevant, and refuses to advocate for keeping cats inside.

It could be conjectured that Audubon, RSPB, and other broad-based conservation organizations have avoided this issue for fear of alienating a portion of their member base. This conjecture is reinforced by *Audubon* magazine's suspension of longtime contributor and editor at large Ted Williams after Williams argued against TNR in an editorial in the *Orlando Sentinel*. Williams ruffled feathers by stating that there were "two humane alternatives to the hell of TNR. One is Tylenol (the human pain medication) —a completely selective feral-cat poison. But the TNR lobby has blocked its registration for this use. The other is trap and euthanize. TE is practiced by state and federal wildlife managers; but municipal TE needs to happen if the annihilation of native wildlife is to be significantly slowed."[13] After much backlash from the conservation community, *Audubon* reinstated Williams, though with limited duties.

🐾

Back at the apartment complex in Oregon's Willamette Valley, Sarah Smith still had no cats in hand after almost an hour of sitting in the van. A few of the residents she had encountered on her arrival approached the vehicle to check on our progress as several other residents stood nearby. The low hum of Interstate 5 could be heard in the near distance. "I went around and talked to everyone about not feeding the cats," the health aide said, returning to the string of the earlier conversation. "They said they weren't. There's one woman who lives with her daughter. I think it's too painful for her to think of the cats going hungry, and she convinces her daughter to feed them."

Well into her second hour, with still no captives, Smith mulled her options. One would be to simply pick up and leave, though she is loath to go home empty-handed. Another would be to leave the traps overnight, though this was not a viable option. "This isn't the kind of place that you can leave a trap," she said. "The crazy people might destroy them, because they think they're hurting the cats. I had two teenage girls destroy my traps at another complex. They jumped up and down on them. At $55 a trap, I don't want to have to replace them. At this point, I don't trust anyone."

A small calico cat with expressive green eyes poked its head up out of the basement. It walked around the traps and sat down. "Come on, sweetie," Smith coaxed. "Go in the trap so you don't get pregnant. Females only have to be four months old to have litters. It's like teenagers getting pregnant—babies having babies." A tawny tabby with brownish stripes made its way across the lawn behind the apartments to the traps. It went from one trap to another, sniffing hard at the food, trying to claw through the top of the trap and then to dig underneath. The tabby stepped halfway into one trap, then backed out. It stuck its head in another, paused, and then stepped in. The trap's door closed, startling the calico. The captured cat circled twice within the cage. Smith stepped briskly from the car, flung a towel over the top of the cage, and placed it in the van. The cat yowled once, rattled around a bit, and became quiet, almost disturbingly so. "These cats are very quiet," Sarah said. "It's part of their survival instinct. I can have a whole carload of cats [she once carried a total of twenty-eight to a clinic in Portland], and it will be silent."

It was nearly dusk, and the apartment complex's shadowy feline population was becoming more active. The calico has been joined by several black cats. The threesome was then joined by a large orange tabby, which promptly began spraying each trap with its scent. A smaller tabby soon joined the group. None seemed too interested in the traps. "A drop trap would be handy," Smith said softly." With a drop trap, the cage is elevated by a stick that is connected to a rope that the trapper can pull when cats are under the cage. Cats tend to be less put off by the elevated cage. There is a slight breeze, which is not helping; the breeze rustles the newspaper, and the cats do not care for that sound. Taking in the five cats, Smith seemed pensive. "It's a shame. Look at these pretty cats. They should be living in someone's home, not under a building. I usually don't wait this long. I probably would've left, except these guys are showing interest. It's supposed to be illegal to kill them, but nobody enforces the law. If people killed dogs like they kill cats, there would be an uproar."

The orange tabby entered a trap with little fanfare, and the door dropped. It circled the trap a few times before Smith draped a towel over the cage and placed it next to the first one. She then fielded a call from someone who reported having a dozen cats roaming her farm. "Do they have any shelter? Good. Are you feeding them? Good. If they have shelter, food and water, those are the main ingredients."

A few more cats were visible, nosing around the cages, just silhouettes in the gloaming. Finally the little calico entered a trap and sprung the door. Smith dropped a towel over the cage and placed it in the car.

It was time to go.

CHAPTER EIGHT

A Landscape with Fewer Free-Ranging Cats: Better for Cats, Better for Birds, Better for People

> The human race is challenged more than ever before to demonstrate our mastery, not over nature but of ourselves.
>
> —Rachel Carson

If you care about animals—especially companion animals—you owe a debt of gratitude to Wayne Pacelle. In his twenty-five years of advocacy leadership at the Humane Society of the United States (HSUS)—including ten years and counting as CEO and president—he has successfully championed many significant animal protection bills and dramatically expanded both the size of the Humane Society (now the 155th largest charity organization in the United States, with annual revenue of $160 million and 11 million members) and the scope of its animal care programs. With movie-star good looks and an Ivy League pedigree, he has been an ideal and effective spokesman for animals. *The NonProfit Times* has named Pacelle to its "Power and Influence Top 50" in five of the last eight years.

As part of his outreach efforts, Pacelle keeps a blog entitled *A Humane Nation*. In several postings over the years, he has spoken to the same question we have been attempting to address in the course of this book: what to do about free-ranging cats? And he (or at least his public relations department) has consistently recognized the paradoxical position in which the issue places an organization like HSUS. In a post from November 2011, Pacelle wrote:

> The Humane Society of the United States advocates for the protection of all animals, and that includes domesticated animals and wildlife. It's often a clear case of right and wrong, and the moral path is clear. There are times, however, when the protection of one species appears to conflict with the protection of another. Perhaps the most common example is the case of outdoor or feral cats and wildlife. Feral cats typically don't live long lives; they're at risk from other cats, dogs, coyotes, cars, disease, and other threats. At the same time, during their lives, they may kill songbirds, small mammals, and other native wildlife, since predation is built into their DNA.[1]

There is no doubt that the scientific community has its differences with Pacelle and HSUS regarding how best to tackle the problems that free-ranging cats pose to the health and well-being of our environment. (For example, HSUS supports trap-neuter-return as an approach for managing cat colonies.) Yet behind the inflammatory rhetoric that frequently cloaks discussions about free-ranging cats, there are many points on which Pacelle, conservationists, and most reasonable parties can agree: (1) there are too many free-ranging cats in America; (2) these cats have an impact on wildlife; (3) cats spread disease to humans; (4) these cats lead a short and often perilous life; (5) domestic cats are better off inside or, at the least, contained; (6) people are the root cause of the problem and as such are morally obligated to address the problem.

It is abundantly clear that free-ranging cats are not the primary threat to the future of birds and other wildlife. Habitat destruction, climate change, and pollution all come to bear on the well-being of wildlife populations; if we as a society hope to maintain these species for future generations, we need to act on all fronts to stem the tide. In the same light, we must act on many different fronts to reduce the populations of free-ranging cats and reduce their impact on native animal populations, both as predators and as vectors of disease. No one solution will prove a silver bullet; only a multipronged strategy will begin to reduce the number of free-ranging cats in the wild. A landscape with no (or at least fewer) free-ranging cats is the only hope for mitigating the toll these animals take on native wildlife and diminishing the spread of disease from cats to human populations.

The first step in working toward a landscape with fewer free-ranging cats is to foster more responsible cat ownership. If pet owners keep abandoning their cats, it will be impossible to ever stem populations of free-ranging animals. If pet owners fail to spay or neuter their cats, it will be impossible to ever stem populations of free-ranging animals. If pet owners continue to let their cats wander freely outside, those cats will take a toll on birds and other wildlife and be subjected to the threats of life on the street—disease, predation, and automobiles, among others.

Convincing people to change their behavior is no easy task; just ask American advertisers, who spent an estimated $177 billion in 2014 to sway people's actions. But this is perhaps the most important step that needs to be taken. Considering cat abandonment, wildlife veterinarian David Jessup drew an analogy with littering. "When I was a four- or five-year-old boy, I'd walk along the road looking for pop bottles to return. There was garbage all over. You could watch people throw crap out the window as they drove by. Over time, attitudes about littering changed, thanks in part to government-sponsored ad campaigns. Now it's considered as disgusting as spitting on the floor in a public building. We need to make abandoning your pet, and failing to get it fixed, and letting it go outside—as socially unacceptable as littering."[2]

One explanation for why so many people have such cavalier attitudes about cats is that they consider the animals somehow able to fend for themselves in the wild—and that being left in a parking lot or park gives the animals a better chance at survival than being dropped off at the local humane society. It may also follow from an assumption that given the surplus of available cats, one animal—should it become slightly inconvenient, given its personality or the owner's life situation—can easily be exchanged for another down the line. Perhaps cats are perceived by some people as relatively expendable. "I think that as a society, we don't value cats," Sharon Harmon, president of the Oregon Humane Society, offered. "If you value an animal, you don't allow it to go without vet care, you bring it in at night, you go look for it if it's lost instead of just getting another one. If you and I were to go to dinner and saw a dog by a dumpster, odds are good we'd try to save it. It doesn't matter how good the restaurant, how long we've waited to get reservations. We're going to try to catch him, check for missing dog notices, take it to a vet. The evening will be gone. If it were a cat by the dumpster, we're not going to miss a beat of conversation. We accept it's the animal's place. Cats aren't served by this attitude."[3]

Anecdotal evidence suggests that a number of pet owners abandon their cats because they believe a visit to an animal shelter is a death sentence, and a loathsome end at that. Taking an animal to the shelter may, for some, carry more shame than abandoning the animal.

If we were to raise the status of animal shelters in the eyes of pet owners, we might discourage some pet owners who are no longer able or willing to care for their cats from leaving them at the city park or college campus. Shelters should be viewed as a resource for the community, a place where your cat has the best chance to find an everlasting home—and they should be funded as such.

It is an unfortunate fact that shelters must kill millions of cats each year; there is simply no room, or resources, to accommodate the tens of millions of unowned/unwanted animals that exist on the landscape. Some will point to so-called no-kill shelters as an alternative to such large-scale euthanasia. While noble in principle, no-kill shelters are not so sanguine in practice. Such shelters have finite resources. Most operate at or near capacity most of the year, which means they must turn away many animals. Some of these animals may then be taken to a traditional shelter, where they will likely be put to sleep. In the case of cats, many owners will simply release their animals onto the landscape, where they frequently face a more prolonged, painful death, while perpetuating the ecological and public-health problems detailed above. A recent trend among no-kill shelters that are trying to enhance their live-release rates is a practice that has been dubbed Return to Field (RTF). Under this model, cats and other stray animals that are picked up by animal control officers (and presumably not part of a colony) are given vaccinations, sterilized, and returned to where they were found. Proponents claim that this gives these "owned but lost" cats a better chance of finding their way home, though how it has been determined that these are "owned" cats is unclear, as they lack microchips or tags. More critical observers might look at RTF as a way for no-kill shelters to pad their live-release numbers. The animals that are returned to the field are likely to die, but not in a humane manner—and not on the no-kill shelter's spreadsheet.

Allowing owned cats to roam freely outside can be seen as another example of irresponsible pet ownership. The advantages of keeping cats inside have been presented here at length. Inside cats are shielded from diseases that can be contracted from other free-ranging cats and wildlife. They are safe from predation by Coyotes, Bobcats, and dogs. They will not be hit by cars. They will not prey on other animals. (Bibs, bells, and other so-called predation deterrents have not been shown to be effective in preventing cats from killing wildlife and do nothing to curb the problem of cats spreading or getting diseases.) And, finally, they are much less likely to transmit diseases to the broader public. Yet in survey after survey, a majority of Americans show a preference for allowing owned cats to wander freely. Part of this potentially destructive attitude seems to stem from ignorance—at times willful—about the impacts individual cats can have on wildlife and about the dangers that await them. ("I understand that some cats might be naughty, but *my* kitty certainly doesn't kill birds.") Part of it stems from an informed desire to let cats be cats; many pet owners believe cats are hardwired to roam and hunt and take the attitude that, as a loving cat owner, "I'm going to enable my cat to pursue this behavior to enrich its life."

This attitude of indifference toward the predatory impact of pet cats on bird and mammal life was confirmed in a 2015 study by Jennifer McDonald and colleagues. McDonald et al. enlisted a number of cat owners in England (in Mawnan Smith, Cornwall) and Scotland (in Thornhill, northwest of Stirling) in a survey about the predatory behavior of their cats. The researchers hoped to determine cat owners' attitudes toward the ecological impacts of domestic cats, to gain insight into potential control strategies, and to learn the influence the predatory behavior of their own cat(s) had on the owners' responses. Thirty-three of the forty-three cats monitored over four months in Mawnan Smith brought a mean of 1.89 animals home each month; ten of the cats did not return home with prey. Twenty-eight of the forty-three cats in Thornhill brought home a mean of 0.81 animals per month over thirteen months surveyed; fifteen cats did not return home with prey. Presented with evidence of their cats' handiwork, 98 percent of the cat owners still were opposed to keeping their cats inside at all times. Sixty percent disagreed that cats were harming wildlife. McDonald et al. concluded that the cat owners in their study failed to perceive the ecological footprint of their cat, rejected the proposition that cats are a threat to wildlife, and opposed management strategies, with the exception of neutering. Perhaps the owners did not view the birds and mammals that fall prey to domestic cats as sentient beings but instead as playthings for their beloved companions.

The American Bird Conservancy has consistently cajoled cat owners to keep their animals inside. For over a decade, ABC has championed its Cats Indoors program to educate the public and policy makers about the many benefits to birds, cats, and people when cats are maintained indoors or under an owner's direct control. The Cats Indoors advertising campaign includes television public service announcements (PSAs), print advertisements, and brochures. One of the PSAs shows a Northern Cardinal at a bird feeder and then shifts to a cat being released from the house, as the owner says "Have fun, Tiger." The cat moves into the yard to a point below the bird feeder as white text appears on the screen: "Cats Kill 2 Billion Birds Each Year" and then "Please Keep Cats Indoors: Better for Cats, Better for Birds, Better for People." Grant

Sizemore, ABC's director of invasive species programs, says, "The idea with the campaign has been to engage different stakeholders about the challenge. We wanted to reach people who worry about cats, people who are interested in wildlife who might not know about the impacts cats have, and people who might not have thought about the issue at all."[4] ABC has distributed more than 100,000 Cats Indoors brochures to interested parties, and the PSAs have had hundreds of airings. But this is hardly the kind of media exposure that will reach the 48 million households that claim cats as pets.

Until recently, the Humane Society of the United States promoted a "cats indoors" policy for owned cats. It asked visitors to its website to sign a pledge, promising to keep their cats indoors for the safety of both cats and wildlife. This page has been taken down. Given the Humane Society's significant membership and the clout it wields in animal welfare circles, it would seem that the organization could do much more.

Veterinarians, pet food manufacturers, and pet supply retailers are three channels that pose great potential for outreach to America's pet owners. Veterinary practices could easily display Cats Indoors posters and brochures in waiting rooms and in consulting rooms, and the vets themselves could deliver a very brief overview on the importance of keeping cats inside at the end of each consultation. Likewise, Procter & Gamble (Iams), Nestlé (Purina, Friskies) Mars (Pedigree, Whiskas), and the other pet-food conglomerates of the world could reach almost every cat owner with a brief Cats Indoors message on cans and bags—and these brands could garner favorable public relations in the process for their efforts to protect cats and wildlife. Major pet supply retailers could follow suit with Cats Indoors messaging in product displays and could print messages on shopping bags. Such efforts will not change behavior overnight, but the power of any advertising rests in repetition and reinforcement. There is little question that the U.S. surgeon general's warning that began appearing on packs of cigarettes in the mid-1960s helped raise the awareness of the link between smoking and cancer, and also helped make smoking seem less socially desirable. Such is the power of on-package information, though it must be noted that this posting was mandated by Congress, not initiated by manufacturers.

Millions of cat owners maintain indoor cats that live long, happy lives. Cats that are not allowed to roam freely still need stimulation, of course, and pet owners have many options. Cat owners can get a leash and walk their cats as tens of millions walk their dogs. Cats can be engaged in the living room with a feather toy or laser pen. There is a new device on the market called One Fast Cat that is essentially a hamster wheel for felines. If cat owners have a home with some outdoor space, they might consider building a "catio" (or cat patio), an outdoor enclosure that allows cats to enjoy fresh air and sunshine while limiting exposure to other animals and automobiles (fig. 8.1). Each fall, the Audubon Society of Portland and the Feral Cat Coalition of Oregon promote a "catio" tour that showcases various enclosure designs in homes around the city.

City, county, state, and federal governments also have a role to play in coaxing American pet owners' toward taking more responsibility for their cats. A number of groups have proposed that municipalities and counties require mandatory licensing for cats. Dogs have been issued licenses in America since at least the 1840s. In 1894 the city of New York passed an ordinance requiring dog owners to obtain a permit for their animal, and most municipalities in America have since followed suit. Licensing, now often bundled with the insertion of a microchip, has many benefits. It allows cities to mandate vaccinations and monitor pet owner compliance; it facilitates reuniting lost animals with their owners; it allows cities and counties to track the number of dogs in their region, both for safety and health reasons; and it provides authorities with a means of distinguishing unowned free-roaming animals from pets. Unowned pets can then be processed according to management policies. (Local ordinances requiring owners to keep pets under control are generally enforced for dogs but not for cats; licensing will facilitate enforcement.)

Despite the widespread acceptance of licensing for dogs, only a small percentage of municipalities require licensing of cats. Why has there been so little interest in applying canine standards to felines? "I believe that with roaming dogs, there has always been a fear factor," ventured Christopher Lepczyk, an associate professor in the School of Forestry and Wildlife Sciences at Auburn University in Alabama, who has conducted extensive research on the impact of cats on wildlife and the efficacy of TNR programs. "Dogs posed a risk to human health from either biting or rabies. That element of fear is not there with cats, though there is the potential for disease. I think we'd all benefit if we began viewing pet ownership as a privilege, not a right."[5] Licensing could be part of the exchange for the privilege. A modest licensing fee of $20 a year per cat could create a windfall of $1.6 billion, money that could be applied toward efforts to reduce populations of free-ranging cats.

The historical role of the cat as part outside worker, part domestic pet, may have also impeded licensing efforts. "Dogs used to be kept outside, but there was a point in time where we started to recognize them as part of the family, and they were brought indoors," Grant Sizemore from ABC says. "We're not completely there with cats. Many people still think of them as barnyard animals that go out and hunt mice. There's a perception that they don't require or even want human interaction or investment. That perception has to change."[6]

As part of any licensing/microchip ordinances, it should be mandatory that cats be spayed or neutered. This is an area in which American pet owners are doing well—according to the Humane Society of the United States, 91 percent of owned cats are sterilized. But if a licensing structure can be universally adopted, there is no reason that a near 100 percent rate cannot be achieved, especially if the cost of the procedure can be subsidized so that it is affordable for all citizens. Given the millions of dollars made available for TNR by charities like Maddie's Fund and PetSmart Charities, funding should not be a problem.

While individual pet owners must take responsibility for their cats, it is everyone's responsibility to address the challenges of managing free-ranging cats to limit their impact on wildlife, both as predators and vectors of disease. From a conservation ecology perspective, the most desirable solution seems clear—remove all free-ranging cats from the landscape by any means necessary. But such a solution is hardly practical given the legions of cats roaming the land—as many as 100 million unowned animals, plus 50 million owned cats that roam—and the painful question of what to do with the cats even if they could be captured. And as we have seen, leaders see little political capital to be gained from stepping into the fray to propose substantive solutions. Faced with budget deficits, affordable housing shortages, and an ever-widening chasm between haves and have-nots, among other woes, leaders rank the plight of wildlife and the misunderstood and poorly known public-health hazards posed by free-ranging cats low on their priority lists, if these issues make the list at all. But given the devastating consequences of free-ranging cats, this needs to change.

By and large, the general public is blissfully unaware of the issue of free-ranging cats. This is in large part because most conservation and bird organizations, wildlife advocates, and conservation ecologists have not been effective at describing the scope of the problem and the scientific facts that speak to definitive action. Take the example of TNR programs. The casually engaged citizen, upon Googling "TNR," would come away with the perception that it is an effective means of managing cat populations, is good for cats, and is in fact practiced by hundreds of municipalities and condoned by such hallowed animal welfare institutions as the Humane Society of the United States. Buried much deeper in the results page—perhaps even on page two or three—there might be links to articles detailing the other side of the story, such as the deficiencies of TNR as a means of population control.

The story of the ecological impact of free-ranging cats is not being heard. Frequently it is being drowned out by the strident and inaccurate claims of free-ranging cat advocates (however well meaning). A cursory review of a leading outdoor-cat advocate's website illustrates some of these claims (presented here with a scientific rebuttal for each):

CAT ADVOCATE CLAIMS: Cats have lived outdoors for more than 10,000 years—they are a natural part of the landscape.

SCIENCE SAYS: Domestic cats are an invasive species throughout their current range, including North America.

CAT ADVOCATE CLAIMS: Today, they live healthy lives outdoors and play important roles in the ecosystem.

SCIENCE SAYS: Free-ranging domestic cats live relatively short, generally brutal lives and play a destructive role in all ecosystems, preying upon and in some cases causing extinctions of native species and acting as vectors of disease.

CAT ADVOCATE CLAIMS: Because TNR is proven to stabilize and reduce cat populations over time, it is now the gold standard for feral cat management in the United States.

SCIENCE SAYS: TNR has not been proven to stabilize and reduce cat populations; in fact, in some cases it has been shown to increase the size of existing colonies (because colonies attract non-neutered and non-spayed cats—including those abandoned in these areas by negligent pet owners).

Thanks to aggressive lobbying efforts, TNR has regrettably become a de facto option for feral cat management; it is not a "gold standard."

One of the most commonly practiced (and effective) tactics deployed by leading free-ranging cat advocacy organizations is to relentlessly attack any studies that cast doubt upon the efficacy of TNR or that disclose the dangers that cats pose to native species and public health, often relying on a flexible interpretation of facts if not outright prevarications. Science organizations rarely respond in similar fashion. The fallout from such misinformation is wide-ranging and severe. It is also testament to the power of perseverance and stridency; call a spoon a knife long enough and loudly enough and people will begin to believe it—or at least they will cease to doubt it and move on to something else.

Wildlife advocates, conservation scientists, and others who care about finding solutions need to make research results known.

Policies promoting cat colonies and TNR should be challenged, and the impacts free-ranging cats have on bird, mammal, and reptile populations must be more clearly illustrated. The urgency of the problem—especially as it pertains to species that are in danger of extirpation or extinction—must be explained. The public must also be alerted to the very real health hazards free-ranging cats pose. The trenches where many of these public opinion battles are fought—local hearings, for example, or outreach events—are open to all. Members of the conservation and science community must mobilize and be heard at such assemblages, even though advocacy is not the traditional role for scientists or a role in which many feel comfortable.

Outdoor-cat advocates claim that nearly 300 cities and counties have enacted ordinances and policies supporting TNR. As of this writing, battle lines over endorsing or rejecting TNR are being drawn in places such as Washington, DC, Delaware, and Sparks, Nevada. For those advocating against TNR programs, it is an uphill battle, but voices for more responsible action can prevail. One recent success story comes from Florida. In the winter of 2013, ABC joined the debate in Florida to combat a proposed state law (Senate Bill 1320) that would have exempted TNR programs from consideration under ordinances prohibiting animal abandonment. Essentially the law would have legalized TNR in the Sunshine State. Working with Audubon Florida, Florida Defenders of Wildlife, People for the Ethical Treatment of Animals, and the Florida Veterinary Medical Association, ABC presented data regarding the prevalence of toxoplasmosis among free-ranging cats and the impact of cats on wildlife. Though the Florida House of Representatives had passed a version of the bill, the Senate version died in the Agriculture Committee. Farther north, in New York, Governor Andrew Cuomo recently vetoed legislation (A2778/S) that would have allocated up to 20 percent of the state's Animal Population Control Program Fund to trap-neuter-return programs, citing evidence that shows TNR does not reduce feral cat populations and that feral cats have a major impact on wildlife. It is unlikely the governor would have taken this stance without pressure from a coalition of stakeholders that included ABC, Audubon New York,

Cornell Lab of Ornithology, birders, animal welfare organizations, and sportsmen's groups.

Government entities can take proactive steps to combat the escalation of free-ranging cat populations. County, state, and federal jurisdictions should enact legislation to ban free-ranging cats from public lands. With such ordinances in place, animal control officers will have legal recourse to remove free-ranging cats that pose a health risk (e.g., cats frequenting a playground in a state or county park) or have the potential to harm wildlife (e.g., cats living near a campground on National Recreation Area land that is near endangered bird nesting sites). There is precedent for such legislation. Title 36 CFR section 2.15 (from the Code of Federal Regulations) states prohibitions for pets in designated National Park Service areas and is summarized as follows:

> Pets or feral animals that are running-at-large and observed by an authorized person in the act of killing, injuring or molesting humans, livestock, or wildlife may be destroyed if necessary for public safety or protection of wildlife, livestock, or other park resources. Pets that do not pose a direct risk to wildlife may be impounded.[7]

While the conservation and public-health communities must confront claims that have potential to steer management policies away from scientific results, they must also maintain lines of communication with such cat advocacy groups, as well as animal shelters and the veterinary community, in an effort to find some common ground—such as increased efforts to spay and neuter and overall animal welfare. A good starting point would be a frank discussion about the fact that cats kill animals. "We'll start to make real progress when cat people accept that cats have an impact," Sharon Harmon from the Oregon Humane Society posited. "Every time a study comes out that says cats have an impact, cat people say 'No, it's humans that do more damage.' We need to agree that cats have a real impact. Until we fully grasp our responsibility for the welfare of all animals—wild and domestic—we're going to have animal mortality and harsh relationships between people who advocate for one or the other."[8]

Over the last decade an unlikely partnership has developed in Portland, Oregon, between two organizations that would seem to be diametrically opposed—the Portland Audubon Society and the Feral Cat Coalition of Oregon. "Over the last ten years, we've established a high level of trust with the Feral Cat Coalition," said Bob Sallinger, Portland Audubon's conservation director. "We work very closely together toward what we feel is a shared goal. People assume it's a tense relationship, but it's not at all. In fact, I'd say it's as close as the working relationships we have with some of our closest conservation partners."[9]

In addition to the catio tour (mentioned earlier), the two groups have collaborated on the Hayden Island Cat Project, a multiyear endeavor to reduce the free-ranging cat population on an islet that sits in the Columbia River between Portland and Vancouver, Washington, that is part of Portland Audubon's larger Cats Safe at Home campaign. Some of the island has been given over to industrial use, though several hundred acres remain undeveloped and provide habitat for some of the area's 200 bird species. Hayden Island has also been used as a dumping ground for unwanted cats and has thus fostered a vibrant free-ranging cat community. The Hayden Island Cat Project aims to assess potential solutions for reducing cat populations on the island, including TNR. "Some in the feral cat community see an effort like this an excuse for capturing and killing cats," Sallinger says, "but that's not what's happening. I think the Feral Cat Coalition people understand that we're trying to work for the general welfare of all animals. Some people in the bird community—here, and in other places across the country—say that we've waved the white flag by advocating for TNR. That somehow we've conceded. I believe that we're working directly on the problem. I'm not saying that we know the best way to handle the cat predation problem here in Portland. But we're trying something different by collaborating with the Feral Cat Coalition in a fully transparent way."

Another example of parties with different beliefs joining together to address a common problem comes from the Hawaiian

island of Kauai. Expressing concern over the burgeoning populations of free-ranging cats, county officials called for a task force
to consider management options and make recommendations. In
2013 the Kauai Feral Cat Task Force was formed, with county
funding. The task force drew its members from the spectrum of interested parties, including representatives from the Kauai Humane
Society, the Forestry and Wildlife division of the State Department
of Land and Natural Resources, the U.S. Fish and Wildlife Service,
local businesses, and the Kauai Albatross Network. In a report
issued in 2014, the Kauai Feral Cat Task Force made eleven recommendations, which include:

- Pass an animal control ordinance that sets a goal of zero
  free-ranging cats by 2025
- Strengthen existing cat licensing ordinances
- Identify sensitive wildlife and cultural areas to guide cat
  enforcement issues at these locales
- Prohibit feeding, sheltering, or maintaining cats on county
  properties
- Make sterilization mandatory for any cats allowed
  outdoors
- Mandate more rigorously managed TNR colonies (or,
  using the task force's wording, "Trap, Neuter, Return &
  Manage colonies"), including 90 percent minimum spay/
  neuter rate
- Make sensitive wildlife and cultural areas high priority
  for trapping of cats
- Initiate (and fund) a public education program[10]

Thus far, the county of Kauai has not implemented any of the task
force's recommendations. Individuals from the Kauai Community
Cat Foundation continue to block implementation by threatening
litigation and deliberate obfuscation. Several species on the U.S.
endangered species list, including the Newell's Shearwater, the Hawaiian Petrel, and the Hawaiian Duck, breed on Kauai and are
preyed upon by cats maintained in colonies.

Though the scientific data make it impossible to advocate for TNR, it must be acknowledged that the practice has become as much a part of the landscape as the cats it purports to help. If TNR and the presence of cat colonies must persist, it is essential that the animals in colonies be truly managed—and that several additional criteria be required. Given the human health risks and wildlife impacts colonies pose, colony sites must be located away from critical wildlife habitats (as determined by wildlife management professionals) and densely populated human settlements. Neutering, vaccinating, and ear-clipping the cats should be only the beginning; at the least, the cats should be implanted with microchips and monitored with automated devices, collared with a license tag, and regularly counted in a standardized fashion by trained personnel at and around colony sites. Because of disease risk, cats should be captured multiple times so vets can administer booster vaccinations and assess health status. Colony caregivers should receive formal training and accreditation to show that they possess a baseline understanding of cat behavior, cat health, and the impact that cats can have on other animals. Municipalities that are condoning and/or underwriting cat colonies also need to monitor their success, just as they monitor any other city program. (Success in this case is defined as a reduction in or extirpation of the overall cat population by a certain date.) City administrators should be mandated to ensure that research is conducted at a subset of colonies within the municipality to better understand the impact of the colony on local wildlife. All colonies must also be adaptively managed; if population reduction goals are not met, or wildlife impacts or disease transmissions are recorded, new strategies or policies must be adopted, including the complete removal of the cats. Penalties should also be imposed upon those individuals who feed free-ranging cats without having the animals spayed or neutered, licensed, and properly managed as described. (No data exist to quantify how many people fall into this category, but it is likely tens of thousands.) "Some caretakers want to claim a level of ownership when it's proposed that a cat be removed," Grant Sizemore from ABC said. "Yet when that same cat bites someone or causes property damage or nuisance, their response is, 'It's not my cat, I just feed it sometimes.' They don't want to take responsibility."[11]

Though it has been stressed again and again, we cannot over-emphasize the importance of spaying and neutering as many cats as possible. Some estimates (cited earlier) place the percentage of a colony's free-ranging cats that have been desexed as low as 2 percent. "We simply can't have cats reproducing at current levels and hope to curtail the number of cats out in the environment," wildlife veterinarian David Jessup says. "As long as we don't reach a higher percentage of spay/neuter, we're on an exponentially ex-panding curve, and we'll never get ahead."[12]

Given the arguments that we have presented thus far in this book, it should be no surprise to the reader to learn that we would find it preferential—if not quite realistic—to see all free-ranging cats removed from the environment. Contrary to what some op-ponents to removal strategies might think, conservationists, ecolo-gists, and public-health officials who desire cats off the landscape *do not wish* to see the animals killed. They simply want them *off the landscape*. In a more perfect world, these animals would be acculturated to human contact whenever possible and adopted out. If assimilation into a human home proved impossible, one option is for animals to be diverted to a cat sanctuary. You may have heard of or visited wild animal sanctuaries—large, fenced-in expanses of land where discarded Lions, elephants, and other "ex-otic" animals can live out their days, with food and any necessary medical care provided by human caregivers. Cat sanctuaries op-erate in a similar manner, accordingly scaled. One such sanctuary is the Cat House on the Kings in central California, near Fresno. This sanctuary provides a home for more than 700 cats, which can wander over a twelve-acre area that is completely enclosed by a cat-proof perimeter fence. There are a number of outbuildings on the property where its residents can take shelter from the ele-ments. When cats are admitted to the Cat House, they are spayed or neutered and receive vaccinations. All eligible cats are available for adoption, and nearly 500 animals are placed in homes each year. Many of the cats living at the Cat House were rescued from the surrounding region; for a fee of $5,000, people who no longer wish to care for their cat can surrender the animal to the Cat House for lifetime support and care.

The efforts of the Cat House and its employees and supporters are laudable. They are helping to remove predators and disease vectors from the environment while providing a safe, nurturing home for the feline residents, whether for a few months or a lifetime. Yet sanctuaries do not appear to be a model that can be scaled to meet the current need. If there are 60 million free-ranging cats (a conservative estimate) in the United States, and we take 700 cats as a population ceiling for an operational sanctuary, we would need nearly 86,000 sanctuaries. If twelve acres is an adequate enclosure size for 700 cats, the 86,000 sanctuaries would take up over 1 million acres, or more than 1,500 square miles—an area larger than Rhode Island. The Cat House estimates that it costs roughly a dollar a day to feed each cat in its care, so the annual food budget alone would approach $22 billion—nearly half the GDP of Rhode Island.

So unless the good citizens of Rhode Island are prepared to vacate their fair state, construct 836 million feet of fence (give or take a few million) before leaving, and relinquish at least half of their earnings to support their feline replacements, sanctuaries may not be a feasible large-scale solution. Nor can we look to conventional community animal shelters to shoulder the burden. According to the ASPCA, the approximately 13,600 independent shelters nationwide take in 3.4 million cats a year; 1.3 million of those cats are adopted, and 1.4 million euthanized. Even if shelters could double or triple their intake and adoption success rates— and this is highly unlikely, given how underfunded most shelters are—these would be mere drops in an ocean of cats.

This all points to a larger realization—that given the scope of the problem and the unique circumstances surrounding each colony—there may not be a one-size-fits-all solution to removing free-ranging cats from the landscape. Even if there were an answer that all parties could agree upon, there would be the question of resources. How would we possibly pay for the trapping of tens of millions of cats, or that fence around Rhode Island? The Department of the Interior does not have the money to underwrite such an undertaking, let alone state fish and game or wildlife departments. But the Centers for Disease Control and Prevention could be another matter. The CDC's projected budget for 2016 is $7 billion; state expenditures for public health amount to over $8.75 billion. Advertising campaigns underwritten by various government entities have helped discourage smoking. Perhaps billboards warning of the potential maladies brought on by cats carrying *Toxoplasma gondii* could similarly impact public behavior. As free-ranging cats become recognized as a serious threat to public health—and thus become viewed as a public-health issue—there is hope that more resources can be devoted to managing the problem.

While large-scale cat removal success stories have thus far proven elusive, there have been a few bright spots. In 1997 Bidwell Park, an appealing 3,670-acre municipal park in Chico, California, that is bifurcated by Big Chico Creek, was becoming overrun with free-ranging cats. The park's proximity to the city made it an easy place for residents to abandon unwanted animals. Bidwell Park was home to a number of bird species, and the California Quail (*Callipepla californica*) populations were plummeting thanks to cat predation. A local conservation organization, Altacal Audubon Society, identified the declining bird species and pressed the city of Chico to take action. Spurred on by Altacal and other community members, the city's Parks and Playground Commission began enforcing Chico's abandonment and litter laws (the latter to prevent individuals from feeding cats living in the park). This discouraged citizens from continuing to abandon their cats in the park but did nothing to remove the animals that were already there. The following year concerned citizens banded together to form the Chico Cat Coalition, with the mission of removing the cats from Bidwell. In the first year, they successfully trapped 440 cats; 340 of the animals were placed into adoption, fifty were sent to live in a barn that served as a sanctuary, and fifty were euthanized. With the considerable reduction in Bidwell's cat population, quail populations rebounded. Ecological balance was restored, with a minimum loss of feline life.

A decade later a more complicated cat removal operation unfolded on San Nicolas Island, roughly sixty miles off the coast of southern California. Cats had been introduced to the thirty-three-square-mile island in the early 1950s, most likely by sailors stationed at the small naval base (to operate a missile telemetry site), and as cats always do, they proliferated. The animals had negatively impacted seabird populations, as well as those of the federally listed threatened Island Night Lizard, Western Snowy Plover, a subspecies of Deer Mouse, and the state-listed threatened Island Fox. Six different organizations (including the U.S. Navy, the U.S. Fish and Wildlife Service, and a nonprofit, Island Conservation) banded together to assemble and implement a plan to remove the cats from San Nicolas—an island perhaps better known as the setting of the acclaimed 1960 children's novel *Island of the Blue Dolphins*. The plan—which deployed 250 padded leg-hold live traps, tracking dogs, and a GIS-enabled trap monitoring system—was implemented over eighteen months. Ultimately fifty-nine cats were captured (unharmed) and flown to the mainland, where they were examined, spayed or neutered, and released to a sanctuary near San Diego provided by the Humane Society of the United States. This program cost over $3 million, or roughly $50,000 per cat.

The cats were removed from San Nicolas Island at a significant cost in large part because they were threatening endangered species. That the animals were trapped and evacuated *alive* may be a function of the navy's resistance to being perceived as the cat-killing branch of the U.S. Armed Forces—and by the availability of ample funds from the Montrose Settlement Restoration Program, a fund set up after millions of pounds of DDTs and PCBs had been discharged into the waters off southern California by Montrose Chemical Corporation. As mentioned earlier, it is paramount—from an ecological as well as a moral perspective—to give endangered species every opportunity to sidestep extinction. As there will not always be deep pockets available to underwrite nonlethal free-ranging cat removal from high-priority wildlife areas, lethal means will have to be considered under certain circumstances.

First, these high-priority regions with habitats that support endangered, threatened, and declining mammals, birds, and other wildlife—and also have cats—must be identified. They are present in every state. A few that come immediately to mind include Cape

May, New Jersey (a key nesting area for Piping Plovers and an important stopover for migrant songbirds); areas around Galveston, Texas (a critical stopover in the United States for Neotropical migrants); areas along the southern shore of Lake Erie, another stopover point for migrants, including endangered species like Kirtland's Warbler and near-threatened species like the Golden-winged Warbler; and the entire island chain of Hawaii (home for a host of endangered species, such as the Hawaiian Monk Seal and thirty-three endangered bird species found nowhere else, including the Hawaiian Duck, Hawaiian Coot, Hawaiian Stilt, and Hawaiian Crow, which is extinct in the wild). In high-priority areas there must be zero tolerance for free-ranging cats. If the animals are trapped, they must be removed from the area and not returned. If homes cannot be found for the animals and no sanctuaries or shelters are available, there is no choice but to euthanize them. If the animals cannot be trapped, other means must be taken to remove them from the landscape—be it the use of select poisons or the retention of professional hunters.

No one likes the idea of killing cats. But sometimes it is necessary. "People need to recognize that by its very definition, euthanization is humane," said David Jessup, the wildlife veterinarian. "It's being put to sleep. It's not something that we want to do wholesale, but it's not an evil thing." The medication that is generally used for euthanasia is pentobarbital, a barbiturate that causes death by respiratory arrest. In appropriate doses, it quickly renders the animal unconscious, shutting down heart and brain functions within one or two minutes.

Some cats will not come to traps, and will need to be removed from the landscape by other means. Australia, as we have seen (chapter 6), is relying upon Curiosity, a sausage-like bait that contains a dose of para-aminopropiophenone, which works by inhibiting the cat's breathing. The bait is distributed by air in the Outback regions where feral cats pose the greatest threat to endemic wildlife, including some of Australia's most threatened species. On Marion Island, in the southern Indian Ocean (1,200 miles south of Cape Town, South Africa), a population of nearly 3,500 cats—progeny of five cats brought to the island in 1949—killed an estimated half million petrels in 1975. To protect petrels and other bird species at risk, the Marion Island Cat Eradication Program was launched in 1977. Several cats were infected with feline panleukopenia, a disease that is easily communicable and ultimately fatal. By 1982 nearly 3,000 cats had been eradicated by contracting panleukopenia. The remaining cats were hunted at night by eight two-man teams using battery-operated spotlights and shotguns, from 1986 through 1989. When hunting ceased to be productive, traps were deployed. By 1991 the eradication project appeared to be complete.

Some who oppose the notion of removing cats from the environment to protect endangered species will cite the inordinate cost of such efforts. The case of Ascension Island in the South Atlantic, part of a British overseas territory, is often trotted out as an example. Seabird populations, which included Sooty Terns (*Onychoprion fuscatus*), Masked Boobies (*Sula dactylatra*), and Ascension Frigatebirds (*Fregata aquila*), were estimated to be a staggering 20 million in 1815 when the British first settled Ascension—bringing with them the island's first cats. Over the next 150 years, nearly all the seabird colonies on the main island were extirpated; relict colonies remained only in cat-inaccessible locations. In 2002 the Royal Society for the Protection of Birds stepped in to lead eradication efforts. Poison baiting and live trapping were the primary means of culling. By 2006 all the cats had been either euthanized or removed; soon after, seabirds began recolonizing sites that had been accessible to cats. By 2012 Ascension Frigatebirds had returned to the island to nest, the first reported nesting in 180 years. The cost of the eradication project was $1.3 million.

Whether you consider $1.3 million an outrageous sum to pay to save a few birds or a wise investment in biodiversity will depend on your philosophical stance. But from a purely financial perspective, there is little question that eradication—at least on a local level—will trump endangered species remediation every time. A breakdown of per species dollars invested in conservation efforts for endangered species from 2004 to 2007 shows that $60.5 million was spent to resuscitate populations of the Southwestern Willow Flycatcher, $67.4 million to protect Red-cockaded Woodpeckers, and nearly $83 million to protect Bald Eagles. (A modest $35.7 million was spent on remediation for Piping Plovers.) Removing free-ranging cats from ecologically sensitive areas is challenging and expensive work, but it is a wise use of resources when compared with the huge investments mandated by the Endangered Species Act once an animal's population has dwindled to dangerously low levels. Investing resources to keep common species common will almost always be more cost-efficient than recovery efforts.

In concert with more draconian steps of cat removal, conservationists and government entities should be prepared to pursue legal action. It has been shown in one court case that releasing cats into the wild and supporting feral cat colonies is a violation of the Migratory Bird Treaty Act and the Endangered Species Act as well as laws prohibiting animal abandonment. It may become necessary to pursue injunctions against colonies and colony managers, particularly in areas that provide habitats for migratory birds or endangered species.

Perhaps the greatest obstacle to convincing humans to take greater responsibility for their pets and act more responsibly on behalf of their environment and the health of the greater society is the growing ignorance and indifference about the natural world. We are an increasingly urban society, increasingly enmeshed with various forms of electronic amusement. Both of these factors serve to reinforce the ever-growing detachment we as a society have developed from the natural world. The further disconnected we become from nature, the less we understand its complexity, its beauty, and, at times, its brutality—and we lose sight of the fact that humans are intricately part of and dependent upon the very natural systems we continue to destroy. If someone does not know about the existence of a Cerulean Warbler, it is unreasonable to think he or she would miss its song once the bird is gone.

A recent poll commissioned by the Nature Conservancy confirmed that American children are spending less time outside and enumerated some of the reasons:

- 80 percent said it was uncomfortable to be outdoors due to things like bugs and heat

- 62 percent said they did not have transportation to natural areas
- 61 percent said there were not natural areas near their homes

There was one bright spot in the poll results. Sixty-six percent of the children surveyed said they had had a personal experience in nature and that it had made them appreciate it more.

In a paper published recently in the *Proceedings of the National Academy of Sciences*, Gregory Bratman and his associates attribute decreased experiences in nature to increased levels of mental illness. Their experiment looked at the impact of nature exposure on rumination, a maladaptive pattern of self-referential thought that is associated with heightened risk for depression and other mental illnesses. Some participants were asked to go on a ninety-minute walk in a natural environment; others took a ninety-minute walk in an urban setting. Those who walked in the natural setting reported lower levels of rumination and showed reduced neural activity in an area of the brain (the subgenual prefrontal cortex) linked to risk for mental illness compared with those who walked through an urban environment. The results suggest that access to nature may be important for maintaining mental health in our increasingly urban world.

Giving more children a chance to experience nature may be the best hope for the wildlife suffering from the proliferation of free-ranging cats on our landscape—and for returning some ecological balance to our environment—because you cannot get people to save something if they do not love it. And you cannot get people to love something if they have not experienced it.

If more Americans had the chance to hold a songbird, look closely into its tiny eyes, and feel its fluttering heart, they might be moved to take action against those threats that imperil these creatures, especially the threats we can control, such as free-ranging cats.

CHAPTER NINE

What Kind of Nature Awaits?

To cherish what remains of the Earth and to foster
its renewal is our only legitimate hope for survival.
—Wendell Berry

Memories of the spring bloom of Washington, DC's beloved cherry
trees were still fresh in the minds of members of the U.S. House
of Representatives as Hugh Hammond Bennett stood before them
on a sunny Friday in May 1934. Bennett, director of the recently
formed Soil Erosion Service, an agency of the U.S. Department of
the Interior, knew something that the congressmen did not: that
a large dust cloud was rapidly moving east from the Great Plains
and would soon deposit millions of tons of dust on the nation's
capital, New York City, Boston, and even on the decks of ships
hundreds of miles off the eastern seaboard. A story in the *New
York Times* reported that airborne topsoil had "lodged itself in
the eyes and throats of weeping and coughing New Yorkers."[1]
Bennett had been conducting soil studies for some thirty years
and had begun warning of the potentially catastrophic effects of
soil erosion in the 1920s—most pointedly in a U.S. Department
of Agriculture study entitled *Soil Erosion: A National Menace*.
Grassland was being plowed under and grazed at unprecedented
and unsustainable rates, leaving no roots to hold the soil together.
With the drought cycle that began in 1931, crops that had been

168

holding the soil intact died, and the topsoil began to literally blow away. Dust storms had been blowing across the plains with increasing frequency, blotting out the sun and enveloping whole structures, yet the gravity of the situation had not struck home with lawmakers or the East Coast establishment until now. As the dark cloud of soil swirled around the Washington Monument, the Lincoln Memorial, and Capitol Hill itself, Bennett is said to have announced to the assembled representatives, "This, gentleman, is what I have been talking about."[2]

Washington got the message and by 1935 had enacted the Soil Conservation Act, mandating crop rotation, new plowing methods, and aggressive grass seeding. In short order, dust storms were reduced by more than half. But these improvements came a little late for the more than 3 million people who had been displaced from Oklahoma, Arkansas, Missouri, Iowa, Nebraska, Kansas, Texas, Colorado, and New Mexico. Acting in ignorance of the consequences of their farming techniques—and in part from blind self-interest (hoping to grow and sell as much wheat as possible)— these Dustbowl refugees were compelled to take to the road in search of a new life. Many experienced the wrenching poverty and humiliation so well captured in the photography of Dorothea Lange and the prose of John Steinbeck's *Grapes of Wrath*.

❧

It is a strange quirk of human nature that we are often unable to respond to problems until they manifest themselves as grand disasters. Whether they are impeded by an unrelenting focus on the short-term, a powerful willingness to filter out bad news, or a tendency toward inertia, many leaders would rather postpone action—take a wait-and-see attitude—than proactively work toward a solution. This tendency appears especially intractable when it comes to environmental or conservation-related issues. For so many generations the American landscape offered seemingly endless bounty—so many bison, so many Passenger Pigeons, so many salmon. Perhaps as a people we have been lulled into thinking that in a land of such plentitude, biological resources are truly infinite and will take care of themselves—until suddenly we are trying to save the last remnant populations or they are already gone.

❧

There is little question that free-ranging cats—both the unowned and the owned pets allowed to roam freely outside—pose a pending ecological and public-health disaster. As of this point, it is equally evident that little if any action is being taken to address this problem; indeed, very few people even *recognize it as a problem*. This raises the question, what sort of emergency—or tragedy—will it take to spur us to action? Will it be the extinction of another species, thanks, in part, to cat predation? Must the Piping Plover, the Hawaiian Petrel, or the Lower Keys Rabbit (or any of a number of other birds, mammals, and other creatures) be lost from the earth forever before we can act? Or will it be an outbreak of disease that has spread from cats to humans—deaths due to rabies or plague, or a spike in cases of schizophrenia and incidents of suicide brought on by the dissemination of *Toxoplasma gondii*?

The consequences of an upsurge in diseases linked to toxoplasmosis are obvious and severe. The fallout we experience as a culture when the populations of a species decline or a species leaves the planet forever is also severe, yet so much harder for many to comprehend. Why does it matter that a small bird or a rodent whose primary habitat is thousands of miles from our home disappears? Nature writer Ted Williams, ruminating on the future of the Florida Grasshopper Sparrow (*Ammodramus savannarum floridanus*), a critically endangered subspecies declining primarily due to habitat loss, speaks to one reason why a species should persevere:

> Maybe the only explanation for people who have to ask why the Florida grasshopper sparrow matters is this: It matters not because it is a source of enrichment for human lives (although it is), not because it is a source of medicine or agent of pest control (it is probably neither), not because it is an "indicator species" that tells us we haven't completely wrecked our habitat, not because it is anything, only because it is.[3]

The novelist and environmentalist Edward Abbey put it slightly differently:

I am weary of the old and tiresome and banal question "Why save the wilderness?" The important and difficult question is "How? How save the wilderness?"[4]

Ecologists will often couch the importance of each species within the notion of the tapestry of life. The understanding here is that every organism has an ecological function that serves the larger ecosystem. Each time a species declines a thread of the tapestry is frayed. Each time a species goes extinct a thread is lost. The more threads that are compromised or lost, the more compromised the tapestry, until finally it unravels. Some of the ecological services provided by species are obvious: honeybees pollinate plants and trees so they can reproduce; birds reduce plant-eating insects so trees can grow; wolves cull Elk that are sick or old so the herd does not exceed the carrying capacity of the land. Others are less obvious. The Gray Wolf (*Canis lupus*), for example, set in motion a cycle of recovery when it was reintroduced into its former territory in the Greater Yellowstone Ecosystem in 1995. After the last wolf in Yellowstone was killed in the 1930s, the population of the Elk (*Cervus elaphus*) slowly ballooned. Additionally, the Elk became less inclined to move about or disperse to avoid predation, and instead focused their browsing—especially in the wintertime—around rivers, where they fed on young willows. The absence of willows suppressed populations of the American Beaver (*Castor canadensis*), which relies on willows to survive. With the return of the wolves Elk stayed on the move and broke into smaller groups, no longer concentrating around the rivers and the willows. The willows recovered, resulting in a return of beavers. Back in healthy numbers, the beavers began building new dams. The dams impacted stream hydrology, helping regulate flow, and created a better rearing habitat—shaded and cooled by the reinvigorated willows—for young Cutthroat Trout (*Oncorhynchus clarki*).

Most ecologists will readily admit that we do not understand the role that each organism plays in the tapestry of life, nor do we exactly comprehend how all the strands fit together. At the same time, they will also express great concern about the future that awaits us should this at once powerful and fragile web begin to unravel.

❧

Birds—and the larger ecosystems they represent—face threats from many directions. Some challenges, like climate change and the loss of habitat brought on by overdevelopment, are so monolithic as to seem insurmountable. One might be willing to leave the car at home and bike to the store to buy a head of organic lettuce, but does this small nod to reducing greenhouse gases make a difference—especially when the lettuce has been trucked in from a farm 200 miles away? The challenge free-ranging cats pose to the well-being of our environment is not on par with climate change. But it is one that each of us can do something about, and it is a problem that can be reversed in a relatively short time. The good news is that nature is resilient once given a chance.

Humans, as a species, also face many threats in the form of virulent disease. Though malaria (caused by a parasite transmitted by mosquitoes) has largely been eradicated in the United States, the disease still claims hundreds of thousands of lives each year, especially in sub-Saharan Africa (627,000 in 2012). We understand the cause of malaria, and we can inoculate people to minimize the probability of disease. If a person becomes infected and is diagnosed early enough, the disease can usually be successfully treated. Emergent and deadly diseases (like Ebola and Zika virus) appear every few years, straining available resources, as epidemiologists, doctors, and public-health officials scramble to understand the disease, to treat those afflicted, and to prevent its spread to the larger population. Some of the various pathogens that free-ranging cats carry and can spread to humans—such as rabies and plague—will likely not reach epidemic proportions. Infection with *Toxoplasma gondii*, however, is already at epidemic proportions and presents us with a different situation entirely. We understand the root cause of this malady, and we have it in our power to avert a looming public-health crisis.

In the previous chapter, we explored some practical solutions for taking control of free-ranging cats. Before we can act upon these recommendations, however, there are two obstacles of a more philosophical nature that stand in our way. The first is the average individual's lack of a grasp on the enormity of the problem. Ecologists spend their time considering and studying concepts of scale—quantifying local data points, say, from a neighborhood, and scaling them up to a city, then regional, then continental level. This issue just is not on the radar of most people. As Jennifer McDonald and her colleagues point out in their study about cat owners' attitudes toward their pets' predation habits (discussed in chapter 8), cat owners need to apprehend how individual predation rates scale up with increased cat densities if they are ever to understand the impacts of cats on wildlife. Without more information and knowledge, it will be difficult for the average person to think about how the handful (or bucketful) of birds killed by his or her cat or those stray cats in the woods behind the local convenience store are emblematic of a widespread and grave problem.

This inability to contextualize scale speaks to a larger challenge— the unwillingness or inability of many people to acknowledge the validity of scientific research, especially if it contradicts their own belief system. *Washington Post* reporter Chris Mooney has studied the topic of science denial extensively. Mooney notes that it is human nature to have blinders on in certain situations and that our prior beliefs have the power to skew how we process new information—they even guide the sort of memories and associations that we summon in our conscious mind. This phenomenon is called confirmation bias. Mooney cites climate-change skepticism as an example of science denial and also the furor over childhood vaccines, which many believe (falsely) to cause autism. In the case of the latter, deniers have created their own media (e.g., the website *Age of Autism*), which in turn are cited as "authorities" and pounce upon any new reports casting doubt on their views with vitriolic critiques and refutations—much in the manner that many outdoor-cat advocacy groups attack any new studies suggesting that free-ranging cats pose a problem.

The second obstacle that needs to be overcome is the unwillingness of some people to acknowledge that euthanasia needs to be part of a successful long-term solution. Michael Soule, a champion of preserving biodiversity who is considered one of the founders of conservation biology sees this unbending adherence to a no-kill ethic as a case of misplaced compassion (fig. 9.1). "There are people in the conservation and animal welfare movement who oppose killing in general," he said recently. "But that's not always the most compassionate stance to take. It would be misplaced compassion not to kill in some instances, as it's the most merciful thing to do in that place and time. But you never forget that killing. It's part of the paradox of compassion. Sometimes you suffer when you are being compassionate."[5]

Many outside of the conservation/ecology community will continue to dismiss the idea that free-ranging cats pose any ecological or public-health danger, no matter what evidence is presented. Some of those willing to acknowledge that cats *might* have an impact on wildlife will hasten to add that the problem is limited to islands. There is no question that free-ranging cats have had (and will continue to have) a disproportionate impact on islands. As previously discussed, many of the confirmed thirty-three cat-driven species extinctions have transpired there. Indeed, many of the endemic birds of the Hawaiian Islands currently reside on the endangered species list, their future hanging in the balance, thanks in part to free-ranging cat predation and disease transmission. But as Stanley Temple's 1989 research showed—and many subsequent studies have confirmed—the impact of free-ranging cats is being felt on mainlands as well, more acutely in some places than others. Temple's research showed that the "islands" of intact grassland in Wisconsin that were maintained as natural habitat between row crops became a focus for cat predation, though they had been established in the first place as a haven for wildlife. Development in urban and suburban areas has resulted in fragmented natural habitats that also resemble islands, albeit islands with a concentration of subsidized predators (e.g., cats who are fed by humans) and an absence of larger predators, like Coyotes. As was discussed earlier, the impact on wildlife in such "islands" is catastrophic. Clearly, cats have impacts on populations on mainland areas. Will free-ranging cat populations take hold on the Great Plains, in the

Sonoran Desert, in the Rockies or Columbia Basin, menacing bird and small mammal populations? It's quite possible. Free-ranging cats—potentially as many as 150 million when owned cats that are allowed outside are included—will likely continue expanding in temperate areas where well-meaning but misguided individuals take pity on them and provide food or in remote areas where there's enough wildlife to sustain them. These cats will continue decimating native wildlife populations and spreading pathogens and the diseases they cause at an accelerated rate as their numbers expand. That is, unless we can garner the political and moral will to begin to take control.

❖

Free-ranging cats are not harbingers of the apocalypse. They will not radically change life as most of us know it in the United States, as climate change and habitat destruction do. But if present trends continue, their growing presence will result in an uptick—perhaps considerable—of zoonotic diseases. And we will continue to see declines in the populations of native birds and other wildlife in regions where free-ranging cats are present. More and more Americans will awake to muted birdsong, if any at all; more and more bird feeders will go unvisited. We will be left with a world that is recognizable but a bit more monochromatic, a bit less diverse.

The state of Florida provides a hint of what may await us, at least from an ecological perspective. Historically, Florida has been one of the most species-rich states in America. In the last century, much of the state's coastal prairie, pine flatwoods, and hardwood hammocks have been cleared for development. Landscapes that have not been cleared—especially in the Everglades—have been impacted by man-made hydrological regimes. Remaining swaths of undeveloped, intact habitat have been bifurcated by roads, creating fragmented "islands" that limit mobility and otherwise compromise ecological functions. These factors—abetted, in some regions, by free-ranging cats—have resulted in Florida having the nation's third-highest number of endangered and threatened species, behind only Hawaii and California: fifty-one federally designated endangered species (including eight birds) and thirty federally designated threatened species (including five birds). Florida is also especially noteworthy for the number of invasive species that have taken hold—more than 500 exotic species of fish and other life-forms have been recorded in the Sunshine State. Some of the better known invasives—the Burmese Python and monitor lizards, for example—have become established mostly because irresponsible pet owners released them into the Everglades (and other wetlands) once they became too large and unruly to handle. The subtropical climate is to their liking, as are the many endemic species they prey on. Should trends continue, Florida's flora and fauna fifty or 100 years down the road will bear only a passing resemblance to the collection of plants and animals there before ground was broken on Disney World.

❧

Not long after the particles of Great Plains dust that enveloped the marble monuments of Washington, DC, in May 1934 were swept away, steps were taken to avert a disaster that might have ultimately decimated America's breadbasket. Americans have also banded together and rallied resources and brainpower to forestall other potential disasters arising from our often uneasy relationship with the natural world. At the dawn of the twentieth century, many people in cities along the eastern seaboard were contracting rabies from roaming packs of dogs. Lawmakers acted to mandate licensure and vaccinations and make it illegal for dogs to roam free. People's attitudes changed about how they treated their dogs, and a legal remedy was created for removing roaming, infected animals from the streets. A similar shift happened with DDT. Thanks to Rachel Carson's work *Silent Spring* we realized that DDT was threatening the future of many bird species—including eagles, falcons, and pelicans. It took ten years and hundreds of millions of dollars, but DDT was eventually banned. And the bird populations bounced back.

❧

Te Papa Tongarewa (National Museum of New Zealand) is an impressive structure sitting above Wellington Harbour along the city's waterfront. With hills in the background, a brewpub across the way, and the taller buildings of the central business district nearby, the setting is slightly reminiscent of San Francisco. Te Papa's six floors contain exhibits dedicated to the natural and cultural history of New Zealand. In addition to the world's largest specimen of the Colossal Squid (1,091 pounds and fourteen feet in length), these include a detailed exhibit on invasive species and their impacts, as well as specimens of New Zealand species that no longer exist. There are models of a Stout-legged Moa, which went extinct in the 1400s due to overhunting by the Maori, and a Haast's Eagle, one of the largest known true raptors, which also went extinct in the 1400s, due to the extinction of its main prey, the Stout-legged Moa. On level three, in a slightly less prominent position than its larger extinct companions, a facsimile model of two Stephens Island Wrens rests inside a small glass case (fig. 9.2). The wrens are here because well-meaning lighthouse keepers brought a cat named Tibbles to an island. The cat—and its progeny—did what cats do: hunt and kill prey. Having evolved without predators, unable to fly and unable to fight back, the Stephens Island Wren was an easy mark. In a matter of a few years it was gone. Now and again, a boy or girl on a field trip who has taken a wrong turn in search of the squid may come upon these small birds—slightly comical-looking, with long beaks, long legs, and slightly ruffled light brown feathers—and read of their demise. He or she may or may not understand what extinction means, or that this is one of only a few examples of this little brown bird left in the world. The wren has been extinct for over 120 years, but it remains a gentle reminder of other species lost forever, thanks to the human introduction of free-ranging cats. Does this make us poorer as a people, as a nation, as a world? We think it does.

Today the actions of well-meaning people and, equally important, the lack of action on the part of others are having the unintended consequences of slowly unraveling the tapestry of our ecological well-being and threatening the health of people around the world. Inside, cats make excellent pets; loose on the landscape, they are—by no fault of their own—unrelenting killers and cauldrons of disease. With cats wandering the landscape it is not difficult to imagine a time in the not-so-distant future when your son or daughter enters a natural history museum and comes upon a small exhibit for a Piping Plover, a Roseate Tern, a Hawaiian Crow, a Florida Scrub Jay, a Key Largo Cotton Mouse, a Choctawhatchee Beach Mouse, a Catalina Island Shrew, a Lower Keys Rabbit, or any number of other species from islands and continents around the world—with the label "Now Extinct."

ACKNOWLEDGMENTS

We are extremely grateful to several people for helping us improve various aspects of the book. For critical content as well as thorough reads of individual chapters we thank Stanley Temple, David Jessup, Robert Marra, and Scott Loss. Critical reviews of the entire book were provided by Chris Lepczyk, Scott Loss, Anne Perrault, Bill Thompson, and Grant Sizemore. Anne Perrault constantly challenged us to make sure we were fair throughout the book. A big thank you to many of our close friends and colleagues who were willing to be sounding boards, including Keith Carlson, Doug Levin, Matt Littlejohn, Ken Matsumoto, Dave Moskowitz, Kevin Omland, Geoff Roach, Janet Rumble, Sonja Scheffer, Scott Sillett, and Tom Will. Our literary agent, Danielle Svetcov, of the Levine Greenberg Rostan Literary Agency, has been a constant source of support on all aspects of the book. Thanks go to Robert Kirk for believing in the potential of this project, as well as to the rest of the team at Princeton University Press, including David Campbell, Mark Bellis, and freelance editor Amy K. Hughes, for helping to push this effort over the finish line. A special thanks also goes out to Tina and Andy Santella for preparing the many pizzas that nurtured our friendship in our early teens. Finally, we thank all animal lovers for their passion and, in many cases, their hard work on behalf of those who don't have a voice in human proceedings. It is our hope that together those who care about animals can navigate a path toward equilibrium for all species.

NOTES

General note to the reader on species names: In works of natural history it is standard practice to capitalize common names of species (e.g., Raccoon, Coyote, Cougar), but not generic names (e.g., skunks, bats, foxes).

CHAPTER ONE: THE OBITUARY OF THE STEPHENS ISLAND WREN

1. There is little record of lighthouse keeper David Lyall's personality, or his thoughts and observations during his stay on Stephens Island in 1894–95. Some of his ruminations and activities above are imagined but consistent with naturalist philosophy and ornithological practices of the time.
2. Medway, "The land bird fauna of Stephens Island."

CHAPTER TWO: AMERICA'S DAIRY LAND AND ITS KILLING FIELDS

1. Brattstrom and Howell, "Birds of the Revilla Gigedo Islands."
2. Stanley Temple, interviewed by Peter Marra, Sep. 24, 2014.

CHAPTER THREE: THE RISE OF BIRD LOVERS AND CAT LOVERS

1. Carlson, *Roger Tory Peterson*, p. 8.
2. Ibid.
3. Seton, *Two Little Savages*, p. 312.
4. Bill Thompson, in discussion with Chris Santella, Nov. 6, 2014.
5. Carlson, *Roger Tory Peterson*, p. 3.
6. Bill Thompson in foreword to Santella, *Fifty Places*.
7. Sharon Harmon, in conversation with Chris Santella, Apr. 10, 2015.
8. Catster, "75 Reasons."
9. State of New York, Article 26.
10. "Cat Colony Caretakers, Episode 3."
11. "Cat Colony Caretakers, Episode 2."
12. "Cat Colony Caretakers, Episode 1."

CHAPTER FOUR: THE SCIENCE OF DECLINE

1. Forbush, *The Domestic Cat*, p. 3.
2. Ibid., p. 29.
3. Ibid., pp. 37–42.
4. Ibid., p. 106.
5. Brinkley, *Wilderness Warrior*, p. 6.
6. Stallcup, "Cats," p. 8.
7. Angier, "That Cuddly Kitty," NYTimes.com.

CHAPTER FIVE: THE ZOMBIE MAKER

1. American Veterinary Medical Association, "AVMA Model Rabies Control Ordinance."
2. Centers for Disease Control and Prevention, "Parasites: Toxoplasmosis."

CHAPTER SIX: TAKING AIM AT THE PROBLEM

1. U.S. Fish and Wildlife Service, "Endangered Species Act | Section 3."
2. U.S. Fish and Wildlife Service, "Digest of Federal Resource Laws."
3. Animal Legal Defense Fund, "Texas."
4. David Favre, in e-mail conversation with Chris Santella, Apr. 21, 2015.
5. Michigan State University College of Law, "Question 62."
6. Beversdorf, *Here, Kitty Kitty*.
7. Stanley Temple, interviewed by Peter Marra, by telephone, Sep. 24, 2014.
8. John Woinarski, interviewed by Chris Santella, by telephone, May 11, 2015.
9. Adams, "Wamsley walks away."
10. Australian Government, Department of the Environment, "Draft."
11. Tharoor, "Australia actually declares 'war'."
12. Gregory Andrews, interviewed by Chris Santella, by email, May 14, 2015.
13. Ramzy, "Australia Writes Morrissey."
14. Cats to Go, https://garethsworld.com/catstogo/#.VvnUzOYp6PU.
15. Gareth Morgan, interviewed by Chris Santella, in person, Dec. 5, 2014.
16. Groc, "Shooting Owls."
17. Bob Sallinger, interviewed by Chris Santella, in person, Apr. 22, 2014.

18. Cornwall, "There Will Be Blood."
19. Marc Beckoff. "U.S. Army Corps of Engineers to Kill Thousands of Cormorants."
20. Barcott, "Kill the Cat."
21. Ibid.
22. CBS News, "Bird Watcher on Trial."
23. Rice, "Galveston Bird Watcher/Cat Killer."
24. Moonraker, "Bird Lovers All Over the World Rejoice."

CHAPTER SEVEN: TRAP-NEUTER-RETURN

1. Sarah Smith, in conversation with Chris Santella, Apr. 10, 2013.
2. People for the Ethical Treatment of Animals, "History."
3. People for the Ethical Treatment of Animals, "What is PETA's stance?"
4. Ron Orchard in conversation with Chris Santella at the Oregon Humane Society, Jul. 21, 2015.
5. Sarah Smith, in conversation with Chris Santella, Apr. 18, 2013.
6. Dauphiné and Cooper, "Impacts of Free-Ranging Domestic Cats," p. 213.
7. City of Houston, "About Trap-Neuter-Return Program."
8. Ibid.
9. Laura Gretch, in discussion with Chris Santella, Apr. 11, 2013.
10. Barrows, "Professional, ethical, and legal dilemmas."
11. Longcore et al., "Critical assessment."
12. National Audubon Society, "National Audubon Society Resolution."
13. Dell'Amore, "Writer's Call to Kill Feral Cats Sparks Outcry."

CHAPTER EIGHT: A LANDSCAPE WITH FEWER FREE-RANGING CATS

1. Pacelle, "Finding Common Ground."
2. David Jessup, in conversation with Chris Santella, Sep. 4, 2015.
3. Sharon Harmon, in conversation with Chris Santella, Apr. 10, 2015.
4. Grant Sizemore, in conversation with Chris Santella, Sep. 9, 2015.
5. Christopher Lepczyk, in conversation with Chris Santella, Sep. 12, 2015.
6. Sizemore, Santella, Sep. 9, 2015.
7. Michigan State University College of Law, "Code of Federal Regulations. Title 36."
8. Harmon, Santella, Apr. 10, 2015.
9. Bob Sallinger, in conversation with Chris Santella, Mar. 15, 2013.

10. Adler, *Kauai Feral Cat Task Force: Final Report.*
11. Sizemore, Santella, Sep. 9, 2015.
12. Jessup, Santella, Sep. 4, 2015.

CHAPTER NINE: WHAT KIND OF NATURE AWAITS?

1. History.com, "This Day in History: May 11, 1934."
2. *American Experience*, "Surviving the Dustbowl."
3. Williams, "The Most Endangered Bird in the Continental U.S."
4. Abbey, "Cactus Chronicles."
5. Michael Soule, in discussion with Chris Santella and Peter Marra, Oct. 17, 2015.

REFERENCES

Abbey, Edward. "Cactus Chronicles." *Orion Magazine*, n.d. https://orionmagazine.org/article/cactus-chronicles/.

Adams, Prue. "Wamsley walks away from Earth Sanctuaries." *Landline*, Mar. 27, 2005. http://www.abc.net.au/landline/content/2005/s1330004.htm.

Adler, Peter S. *Kauai Feral Cat Task Force: Final Report*. Mar. 2014. http://www.accord3.com/docs/FCTF%20Report%20FINAL.pdf.

Alley Cat Allies. "Cats & The Environment Resource Center." http://www.alleycat.org/page.aspx?pid=324 (accessed Dec. 28, 2015).

Alley Cat Allies. "Smithsonian-Funded Junk Science Gets Cats Killed." http://www.alleycat.org/sslpage.aspx?pid=1443 (accessed Dec. 28, 2015).

American Bird Conservancy. "WatchList Species Account for Piping Plover." http://www.abcbirds.org/abcprograms/science/watchlist/piping_plover.html (accessed Jul. 22, 2015).

American Experience. "Surviving the Dustbowl." 2007. http://www.pbs.org/wgbh/americanexperience/features/transcript/dustbowl-transcript/.

American Museum of Natural History. ". . . an on-going process." http://www.amnh.org/science/biodiversity/extinction/Intro/OngoingProcess.html (accessed Apr. 19, 2015).

American Pet Products Association National Pet Owners Survey 2011–2012. Greenwich, CT: American Pet Products Manufacturers Association, Inc., 2011.

American Society for the Prevention of Cruelty to Animals. "Shelter Intake and Surrender." https://www.aspca.org/animal-homelessness/shelter-intake-and-surrender (accessed Jul. 12, 2015).

American Veterinary Medical Association. "AVMA Model Rabies Control Ordinance." https://www.avma.org/KB/Policies/Documents/avma-model-rabies-ordinance.pdf (accessed Sep. 20, 2015).

American Veterinary Medical Association Pet Ownership and Demographics Sourcebook, 2nd ed. Schaumburg, IL: American Veterinary Medical Association, 2007.

Angier, N. "That Cuddly Kitty Is Deadlier Than You Think." *New York Times*, Jan. 29, 2013. http://www.nytimes.com/2013/01/30/science/that-cuddly-kitty-of-yours-is-a-killer.html.

"Animal Equity–YouTube." https://www.youtube.com/user/animalequity.

Animal Legal Defense Fund. "Animal Protection Laws of Texas." In *Animal Protection Laws of the USA and Canada*. 8th ed. 2013. http://aldf.org/wp-content/themes/aldf/compendium-map/us/2013/TEXAS.pdf.

Aramini, J. J., C. Stephen, J. P. Dubey, C. Engelstoft, H. Schwantje, and C. S. Ribble. "Potential contamination of drinking water with *Toxoplasma gondii* oocysts." *Epidemiology and Infection* 122, no. 2 (1999): 305–15.

Australian Government. Department of the Environment. "Draft: Threat abatement plan for predation by feral cats." https://www.environment.gov.au/biodiversity/threatened/threat-abatement-plans/draft-feral-cats-2015 (accessed Jul. 28, 2015).

Baptista, L. F., and J. E. Martínez-Gómez. "El programa de reproducción de la Paloma de la Isla Socorro, *Zenaida graysoni*." *Ciencia y Desarrollo* 22 (1996): 30–35.

Barcott, Bruce. "Kill the Cat That Kills the Bird?" *New York Times Magazine*, Dec. 2, 2007. http://www.nytimes.com/2007/12/02/magazine/02cats-v--birds-t.html?_r=0 (accessed Sep. 1, 2015).

Barnosky, Anthony D., Nicholas Matzke, Susumu Tomiya, Guinevere OU Wogan, Brian Swartz, Tiago B. Quental, Charles Marshall, et al. "Has the Earth's sixth mass extinction already arrived?" *Nature* 471, no. 7336 (2011): 51–57.

Barrows, Paul L. "Professional, ethical, and legal dilemmas of trap-neuter-release." *Journal of the American Veterinary Medical Association* 225 (2004): 1365–69.

Beckoff, Marc. "U.S. Army Corps of Engineers to Kill Thousands of Cormorants: There Will Be Blood." *HuffPost Green*. http://www.huffingtonpost.com/marc-bekoff/u-s-army-corps-of-engineers-to-kill-thousands-of-cormorants-there-will-be-blood_b_6964178.html.

Benenson, Michael W., Ernest T. Takafuji, Stanley M. Lemon, Robert L. Greenup, and Alexander J. Sulzer. "Oocyst-transmitted toxoplasmosis associated with ingestion of contaminated water." *New England Journal of Medicine* 307, no. 11 (1982): 666–69.

Berdoy, M., J. P. Webster, and D. W. Macdonald. "Fatal attraction in *Toxoplasma*-infected rats: A case of parasite manipulation of its mammalian host." In *Proceedings of the Royal Society B*, vol. 267 (2000): 267.

Beversdorf, Andy (dir.). *Here, Kitty Kitty*. Prolefeed Studios, 2007.

Blancher, P. "Estimated Number of Birds Killed by House Cats (*Felis catus*) in Canada / Estimation du nombre d'oiseaux tués par les chats domestiques (*Felis catus*) au Canada." *Avian Conservation and Ecology* 8.2 (2013): 3.

Blanton, J. D., D. Palmer, and C. E. Rupprecht. "Rabies surveillance in the United States during 2009." *Journal of the American Veterinary Medical Association* 237 (2010): 646–57.

Bonnington, C., K. J. Gaston, and K. L. Evans. "Fearing the feline: Domestic cats reduce avian fecundity through trait-mediated indirect effects that increase nest predation by other species." *Journal of Applied Ecology* 40 (2013): 15–24.

Bratman, Gregory N., J. Paul Hamilton, Kevin S. Hahn, Gretchen C. Daily, and James J. Gross. "Nature experience reduces rumination and subgenual prefrontal cortex activation." *Proceedings of the National Academy of Sciences* 112, no. 28 (2015): 8567–72.

Brattstrom, Bayard H., and Thomas R. Howell. "The Birds of the Revilla Gigedo Islands, Mexico." *Condor* 58, no. 2 (1956): 107–20. doi:10.2307/1364977.

Brautigan, Richard. "The Good Work of Chickens." In *The Revenge of the Lawn*. New York: Simon & Schuster, 1971.

Brinkley, Douglas. *The Wilderness Warrior: Theodore Roosevelt and the Crusade for America*. New York: HarperCollins, 2009.

Campagnolo, E. R., L. R. Lind, J. M. Long, M. E. Moll, J. T. Rankin, K. F. Martin, M. P. Deasy, V. M. Dato, and S. M. Ostroff. "Human Exposure to Rabid Free-Ranging Cats: A Continuing Public Health Concern in Pennsylvania." *Zoonoses and Public Health* 61, no. 5 (2014): 346–55.

Carlson, Douglas. *Roger Tory Peterson: A Biography*. Austin: University of Texas Press, 2012.

"Cat Colony Caretakers, Episode 1," Animal Equity. https://www.youtube.com/watch?v=2EMBlNr4CbM.

"Cat Colony Caretakers, Episode 2," Animal Equity. https://www.youtube.com/watch?v=1-9edYnQs5U.

"Cat Colony Caretakers, Episode 3," Animal Equity. https://www.youtube.com/watch?v=mxHLAmLNvSw.

Cat House on the Kings. "What We Do." http://www.cathouseonthekings.com/whatwedo.php (accessed Oct. 12, 2015).

Catster. "75 Reasons to Love Cats." http://www.catster.com/lifestyle/75-reasons-to-love-cats.

Cats to Go website. https://garethsworld.com/catstogo/.

CBS News. "Bird Watcher on Trial for Killing Cat." Nov. 16, 2007. http://www.cbsnews.com/news/bird-watcher-on-trial-for-killing-cat/ (accessed Jul. 18, 2015).

Ceballos, G., P. R. Ehrlich, A. D. Barnosky, A. García, R. M. Pringle, and T. M. Palmer, "Accelerated modern human-induced species losses: Entering the sixth mass extinction." *Science Advances* 1 (2015): e1400253.

Centers for Disease Control and Prevention. "Compendium of Animal Rabies Prevention and Control." *Morbidity and Mortality Weekly Report*, Nov. 4, 2011 (R.R. vol. 60, no. 6): 1–18. http://www.cdc.gov /mmwr/pdf/rr/rr6006.pdf.

Centers for Disease Control and Prevention. "Parasites: Toxoplasmosis (*Toxoplasma* infection)." http://www.cdc.gov/parasites/toxoplasmosis/ (accessed Sep. 19, 2015).

Centers for Disease Control and Prevention. "Parasites: Toxoplasmosis (*Toxoplasma* infection). Biology." http://www.cdc.gov/parasites /toxoplasmosis/biology.html (accessed Sep. 19, 2015).

Centers for Disease Control and Prevention. "Rabies Surveillance in the U.S.: Domestic Animals—Rabies." http://www.cdc.gov/rabies/location /usa/surveillance/domestic_animals.html (accessed Sep. 20, 2015).

Churcher, P. B., and J. H. Lawton. "Predation by domestic cats in an English village." *Journal of Zoology* 212, no. 3 (1987): 439–55.

City of Houston. "About Trap-Neuter-Return Program." http://www .houstontx.gov/barc/trap_neuter_return.html (accessed Aug. 23, 2015).

Coelho, F. M., M. R. Q. Bomfim, F. de Andrade Caxito, N. A. Ribeiro, M. M. Luppi, É. A. Costa, and M. Resende. "Naturally occurring feline leukemia virus subgroup A and B infections in urban domestic cats." *Journal of General Virology* 89, no. 11 (2008): 2799–2805.

Coleman, J. S., and S. A. Temple. "Effects of free-ranging cats on wildlife: A progress report." Fourth Eastern Wildlife Damage Control Conference (1989).

Coleman, John S., and Stanley A. Temple. "Rural residents' free-ranging domestic cats: A survey." *Wildlife Society Bulletin (1973–2006)* 21, no. 4 (1993): 381–90.

Cornell University College of Veterinary Medicine, Cornell Feline Health Center. "Feline Leukemia Virus." http://www.vet.cornell.edu/fhc /health_information/brochure_felv.cfm (accessed Mar. 7, 2015).

Cornwall, Warren. "There Will Be Blood." *Conservation*, Oct. 24, 2014. http://conservationmagazine.org/2014/10/killing-for-conservation/.

Crooks, D. R., and M. E. Soule. "Mesopredator release and avifaunal extinctions in a fragmented system." *Nature* 400 (1999): 563–66.

Cunningham, Mark, Brown, Shindle, Terrell, Hayes, Ferree, McBride, Blankenship, Jansen, Citino, Roelke, Kiltie, Troyer, O'Brien. "Epizootiology and Management of Feline Leukemia Virus in the Florida Puma." *Journal of Wildlife Diseases* 44, no. 3 (July 2008): 537–52. doi:10.7589/0090-3558-44.3.537.

Daniels, M. J., M. C. Golder, O. Jarrett, and D. W. MacDonald. "Feline viruses in wildcats from Scotland." *Journal of Wildlife Diseases* 35, 1 (1999): 121–24.

Dauphiné, Nico, and Robert J. Cooper. "Impacts of Free-Ranging Domestic Cats (*Felis catus*) on Birds in the United States: A Review of Recent Research, with Conservation and Management Recommendations." *Proceedings of the Fourth International Partners in Flight Conference: Tundra to Tropics*, Oct. 2009, pp. 205–19. http://www.partnersinflight .org/pubs/McAllenProc/articles/PIF09_Anthropogenic%20Impacts /Dauphine_1_PIF09.pdf.

Dawson, T. "Cat Disease Threatens Endangered Monk Seals." *Scientific American*, Dec. 7, 2010. http://www.scientificamerican.com/article/cat -disease-threatens-endangered-monk-seals/ (accessed Mar. 17, 2016).

Dell'Amore, Christine. "Writer's Call to Kill Feral Cats Sparks Outcry." *National Geographic*, Mar. 22, 2013. http://news.nationalgeographic .com/news/2013/03/130320-feral-cats-euthanize-ted-williams -audubon-science/.

Doll, J. M., P. S. Seitz, P. Ettestad, A. L. Bucholtz, T. Davis, et al. "Cat transmitted fatal pneumonic plague in a person who travelled from Colorado to Arizona." *American Journal of Tropical Medicine and Hygiene* 51 (1994): 109–14.

Dyer, Jessie L., Ryan Wallace, Lillian Orciari, Dillon Hightower, Pamela Yager, and Jesse D. Blanton. "Rabies surveillance in the United States during 2012." *Journal of the American Veterinary Medical Association* 243, no. 6 (2013): 805–15.

eMarketer. "U.S. Total Media Ad Spend Inches Up, Pushed by Digital." http://www.emarketer.com/Article/US-Total-Media-Ad-Spend-Inches -Up-Pushed-by-Digital/1010154 (accessed Sep. 24, 2015).

Filoni, C., J. L. Catão-Dias, G. Bay, E. L. Durigon, R. S. P. Jorge, H. Lutz, and R. Hofmann-Lehmann. "First evidence of feline herpesvirus, calicivirus, parvovirus, and *Ehrlichia* exposure in Brazilian free-ranging felids." *Journal of Wildlife Diseases* 42, no. 2 (2006): 470–77.

Flegr, Jaroslav. "How and why *Toxoplasma* makes us crazy." *Trends in Parasitology* 29, no. 4 (2013): 156–63.

Flegr, J., J. Prandota, M. Sovičková, and Z. H. Israili. "Toxoplasmosis—a global threat. Correlation of latent toxoplasmosis with specific disease burden in a set of 88 countries." *PLoS One* 9, 3 (Mar. 2014): e90203. doi:10.1371/journal.pone.0090203.

Foley, Patrick, Janet E. Foley, Julie K. Levy, and Terry Paik. "Analysis of the impact of trap-neuter-return programs on populations of feral cats." *Journal of the American Veterinary Medical Association* 227, no. 11 (2005): 1775–81.

Fooks, A. R., A. C. Banyard, D. L. Horton, N. Johnson, L. M. McElhinney, and A. C. Jackson. "Current status of rabies and prospects for elimination." *The Lancet* 384, no. 9951 (2014): 1389–99.

Forbush, E. H. *The Domestic Cat: Bird Killer, Mouser and Destroyer of Wild Life, Means of Utilizing and Controlling It.* Boston, MA: Wright & Potter Printing Co., 1916.

Fromont, E., D. Pontier, A. Sager, F. Leger, F. Bourguemestre, E. Jouquelet, P. Stahl, and M. Artois. "Prevalence and pathogenicity of retroviruses in wildcats in France." *The Veterinary Record* 146, 11 (2000): 317–19.

Gage K. L., D. T. Dennis, K. A. Orloski, P. J. Ettestad, T. L. Brown, et al. "Cases of cat-associated human plague in the Western US, 1977–1998." *Clinical Infectious Diseases* 30 (2000): 893–900.

Galbreath, R., and D. Brown. "The tale of the lighthouse-keeper's cat: Discovery and extinction of the Stephens Island wren (*Traversia lyalli*)." *Notornis* 51, no. 4 (2004): 193–200.

Gratz, N. G. "Rodent reservoirs & flea vectors of natural foci of plague." In O. T. Dennis, K. L. Gage, N. Gratz, J. D. Poland, and E. Tikhomirov (eds.), *Plague Manual: Epidemiology, Distribution, Surveillance and Control*, pp. 63–96. Geneva, Switzerland: World Health Organization, 1999.

Grier, Katherine C. *Pets in America: A History.* Chapel Hill: University of North Carolina Press, 2006.

Groc, Isabelle. "Shooting Owls to Save Other Owls." *National Geographic*, Jul. 19, 2014. http://news.nationalgeographic.com/news/2014/07/140717-spotted-owls-barred-shooting-logging-endangered-species-science/.

Gunther, Idit, Hilit Finkler, and Joseph Terkel. "Demographic differences between urban feeding groups of neutered and sexually intact free-roaming cats following a trap-neuter-return procedure." *Journal of the American Veterinary Medical Association* 238, no. 9 (2011): 1134–40.

Hamilton Raven, Peter, and George Brooks Johnson. *Biology.* New York: McGraw-Hill Education, 2002, p. 68.

Hanson, Chad C., Jake E. Bonham, Karl J. Campbell, Brad S. Keitt, Annie E. Little, and Grace Smith. "The Removal of Feral Cats from San Nicolas Island: Methodology." In R. M. Timm and K. A. Fagerstone (eds.), *Proceedings: 24th Vertebrate Pest Control Conference*, pp. 72–78. Davis: University of California, 2010. http://www.islandconservation.org/UserFiles/File/Hanson%20et%20al%202010_final.pdf (accessed Oct. 12, 2015).

Hayhow, D. B., G. Conway, M. A. Eaton, P. V. Grice, C. Hall, C. A. Holt, A. Kuepfer, D. G. Noble, S. Oppel, K. Risely, C. Stringer, D. A. Stroud, N. Wilkinson, and S. Wotton. *The State of the UK's Birds 2014.* Sandy, Bedfordshire: RSPB, BTO, WWT, JNCC, NE, NIEA, NRW, and SNH, 2014.

Held, J. R., E. S. Tierkel, and J. H. Steele. "Rabies in man and animals in the United States, 1946–65." *Public Health Report* 82 (1967): 1009–18.

History.com. "This Day in History: May 11, 1934: Dust storm sweeps from Great Plains across Eastern states." http://www.history.com/this-day-in-history/dust-storm-sweeps-from-great-plains-across-eastern-states (accessed Nov. 2, 2015).

House, Patrick K., Ajai Vyas, and Robert Sapolsky. "Predator cat odors activate sexual arousal pathways in brains of *Toxoplasma gondii* infected rats." *PLoS One* (2011): e23277.

Houser, Susan. "A New Way to Save Shelter Cats." *HuffPost Impact*, Jan. 4, 2016. http://www.huffingtonpost.com/susan-houser/return-to-field-a-new-con_b_8911786.html (accessed Jan. 4, 2016).

Humane Society of the United States. *The Outdoor Cat: Science and Policy from a Global Perspective.* Marina del Rey, CA, Dec. 3–4, 2012. http://www.humanesociety.org/assets/pdfs/pets/outdoor_cat_white_paper.pdf.

International Union for Conservation of Nature and Natural Resources. *The IUCN Red List of Threatened Species*, vers. 2014.3. http://www. iucnredlist.org.

Izawa, M., T. Doi, and Y. Ono. "Grouping patterns of feral cats (*Felis catus*) living on a small island in Japan." *Japanese Journal of Ecology* 32 (1982): 373–82.

Jehl, J. R., and K. C. Parkes. "Replacements of landbird species on Socorro Islands, Mexico." *The Auk* 100 (1983): 551–59.

Jehl, J. R., and K. C. Parkes. "The status of the avifauna of the Revillagigedo Islands, Mexico." *Wilson Bulletin* 94 (1982): 1–19.

Jessup, David A. "The Welfare of feral cats and wildlife." *Journal of the American Veterinary Association*, vol. 225, no. 9 (Animal Welfare Forum: Management of Abandoned and Feral Cats, 2004): 1377–83.

Jessup, D. A., K. C. Pettan, L. J. Lowenstine, and N. C. Pedersen. "Feline leukemia virus infection and renal spirochetosis in a free-ranging cougar (*Felis concolor*)." *Journal of Zoo and Wildlife Medicine* 24 (1993): 73–79.

Kays, Roland, Robert Costello, Tavis Forrester, Megan C. Baker, Arielle W. Parsons, Elizabeth L. Kalies, George Hess, Joshua J. Millspaugh, and William McShea. "Cats are rare where coyotes roam." *Journal of Mammalogy* 96, no. 5 (2015): 981–87.

Kays, Roland W., and Amielle A. DeWan. "Ecological impact of inside/outside house cats around a suburban nature preserve." *Animal Conservation* 7, no. 3 (2004): 273–83.

Knight, Kathryn. "How pernicious parasites turn victims into zombies." *Journal of Experimental Biology* 216, no. 1 (2013): i–iv.

Kreuder, C., M. A. Miller, D. A. Jessup, L. J. Lowenstine, M. D. Harris, J. A. Ames, T. E. Carpenter, P. A. Conrad, and J. A. K. Mazet. "Patterns of Mortality in Southern Sea Otters (*Enhydra lutris nereis*) from 1998–2001." *Journal of Wildlife Diseases* 39, no. 3 (Jul. 2003): 495–509.

Kunin, W. E., and Kevin Gaston, eds. *The Biology of Rarity: Causes and Consequences of Rare–Common Differences*. Springer Netherlands, 1996.

Lawton, J., and R. May. *Extinction Rates*. Oxford and New York: Oxford University Press, 1995.

Lepczyk, Christopher A., Nico Dauphine, David M. Bird, Sheila Conant, Robert J. Cooper, David C. Duffy, Pamela Jo Hatley, Peter P. Marra, Elizabeth Stone, Stanley A. Temple. "What Conservation Biologists Can Do to Counter Trap-Neuter-Return: Response to Longcore et al." *Conservation Biology* (2010): 1–3.

Levy, J. K., and P. C. Crawford. "Humane strategies for controlling feral cat populations." *Journal of the American Veterinary Medical Association* 225, no. 9 (2004): 1354–60.

Levy, Julie K., David W. Gale, and Leslie A. Gale. "Evaluation of the effect of a long-term trap-neuter-return and adoption program on a free-roaming cat population." *Journal of the American Veterinary Medical Association* 222, no. 1 (2003): 42–46.

Ling, Vinita J., David Lester, Preben Bo Mortensen, Patricia W. Langenberg, and Teodor T. Postolache. "*Toxoplasma gondii* seropositivity and suicide rates in women." *Journal of Nervous and Mental Disease* 199, no. 7 (2011): 440.

LLRX.com. "The Domestic Cat and the Law: A Guide to Available Resources." http://www.llrx.com/features/catlaw.htm.

Lohr, Cheryl A., Christopher A. Lepczyk, and Linda J. Cox. "Identifying people's most preferred management technique for feral cats in Hawaii." *Human–Wildlife Interactions*, no. 8 (2014): 56–66.

Longcore, Travis, Catherine Rich, and Lauren M. Sullivan. "Critical assessment of claims regarding management of feral cats by trap-neuter-return." *Conservation Biology* 23, no. 4 (2009): 887–94.

Loss, Scott R., Tom Will, and Peter P. Marra. "Direct Mortality of Birds from Anthropogenic Causes." *Annual Review of Ecology, Evolution, and Systematics* 46, no. 1 (2015).

Loss, S. R., Tom Will, and Peter P. Marra. "Direct human-caused mortality of birds: Improving quantification of magnitude and assessment of population impacts." *Frontiers in Ecology and Environment* 10 (2012): 357–64.

Loss, S. R., Tom Will, and Peter P. Marra. "The impact of free-ranging domestic cats on wildlife of the United States." *Nature Communications* 4 (2013): 1396.

Lowe, Sarah, Michael Browne, Souyad Boudjelas, and M. De Poorter. *100 of the World's Worst Invasive Alien Species: A Selection from the*

Global Invasive Species Database. Auckland, New Zealand: The Invasive Species Specialist Group, n.d.

Loyd, K.A.T., S. M. Hernandez, J. P. Carroll, K. J. Abernathy, and G. J. Marshall. "Quantifying free-ranging domestic cat predation using animal-borne video cameras." *Biological Conservation* 160 (2013): 183–89.

Martínez, J. E., and R. L. Curry. "Conservation status of the Socorro Mockingbird in 1993–94." *Bird Conservation International* 6 (1996): 271–83.

Martínez-Gómez, J. E., A. Flores-Palacios, and R. L. Curry. "Habitat requirements of the Socorro Mockingbird, *Mimodes graysoni*." *Ibis* 143 (2001): 456–67.

May, John Bichard. *Edward Howe Forbush: A Biographical Sketch*. Edited by Robert F. Cheney. Boston: Society from the William Brewster Fund, 1928.

Maynard, L. W. "President Roosevelt's List of Birds, seen in the White House Grounds and about Washington during his administration." *Bird Lore* 12, no. 2 (1910). http://www.theodore-roosevelt.com/images /research/trbirdswhitehouse.pdf.

McAuliffe, Kathleen. "How Your Cat Is Making You Crazy." *The Atlantic*, Mar. 2012.

McDonald, Jennifer L., Mairead Maclean, Matthew R. Evans, and Dave J. Hodgson. "Reconciling actual and perceived rates of predation by domestic cats." *Ecology and Evolution* 5, no. 14 (2015): 2745–53.

Medina, Félix M., Elsa Bonnaud, Eric Vidal, Bernie R. Tershy, Erika S. Zavaleta, C. Josh Donlan, Bradford S. Keitt, Matthieu Corre, Sarah V. Horwath, and Manuel Nogales. "A global review of the impacts of invasive cats on island endangered vertebrates." *Global Change Biology* 17, no. 11 (2011): 3503–10.

Medway, D. G. "The land bird fauna of Stephens Island, New Zealand in the early 1890s, and the cause of its demise." *Notornis* 51 (2004): 201–11.

Meli, M. L., V. Cattori, F. Martínez, G. López, A. Vargas, M. A. Simón, H. Lutz, et al. Feline leukemia virus and other pathogens as important threats to the survival of the critically endangered Iberian lynx (*Lynx pardinus*)." *PLoS One* 4, 3 (2009): e4744.

Michigan State University College of Law. "Code of Federal Regulations. Title 36." https://www.animallaw.info/administrative/us-dogs-large -part-2-resource-protection-public-use-and-recreation-%C2%A7-215 -pets (accessed Oct. 3, 2015).

Michigan State University College of Law. "Question 62–Feral Cats– DEFEATED." https://www.animallaw.info/statute/wi-cats-question-62 -defeated.

Millán, J., and A. Rodríguez. "A serological survey of common feline pathogens in free-living European wildcats (*Felis silvestris*) in central Spain." *European Journal of Wildlife Research* 55, 3 (2009): 285–91.

Millener, P. R. "The only flightless passerine: The Stephens Island Wren (*Traversia lyalli*: Acanthisittidae)." *Notornis* 36, 4 (1989): 280–84

Modern Cat. "TNR Week: A Brief History of TNR—Q&A with Ellen Perry Berkeley." http://www.moderncat.net/2010/09/14/tnr-week-a-brief-history-of-tnr-qa-with-ellen-perry-berkeley/ (accessed Sep. 5, 2015).

Mooney, Chris. "The Science of Why We Don't Believe Science." *Mother Jones*, May/June 2011. http://www.motherjones.com/politics/2011/03/denial-science-chris-mooney (accessed Oct. 8, 2015).

Moonraker. "Bird Lovers All Over the World Rejoice as Serial Killer James M. Stevenson Is Rewarded by a Galveston Court for Gunning Down Hundreds of Cats." *Cat Defender*, Nov. 20, 2007. http://catdefender.blogspot.com/2007/11/bird-lovers-all-over-world-rejoice-as.html.

Moseby, K. E., and B. M. Hill. "The use of poison baits to control feral cats and red foxes in arid South Australia. I. Aerial baiting trials." *Wildlife Research* 38, 4 (2011): 338–50.

Moura, Lenildo D. E., Lilian Maria Garcia Bahia Oliveira, Marcelo Yoshito Wada, Jeffrey L. Jones, Suely Hiromi Tuboi, Eduardo H. Carmo, Walter Massa Ramalho, et al. "Waterborne toxoplasmosis, Brazil, from field to gene." *Emerging Infectious Diseases*, vol. 12, no. 2 (Feb. 2006): 326–29. http://wwwnc.cdc.gov/eid/article/12/2/pdfs/04-1115.pdf.

National Audubon Society. "About Us." https://www.audubon.org/about.

National Audubon Society. "Beating the Odds: A Year in the Life of a Piping Plover." http://docs.audubon.org/plover (accessed Jul. 20, 2015).

National Audubon Society. "National Audubon Society Resolution: Resolution Approved by the Board of Directors on Dec. 7, 1997, Regarding Control and Management of Feral and Free-Ranging Domestic Cats." http://web4.audubon.org/local/cn/98march/nasr.html (accessed Sep. 21, 2015).

National Oceanic and Atmospheric Administration, Montrose Settlements Restoration Program. "About Us." http://www.montroserestoration.noaa.gov/about-us/ (accessed Oct. 13, 2015).

National Park Service. "Spotted Owl and Barred Owl." http://www.nps.gov/redw/learn/nature/spotted-owl-and-barred-owl.htm (accessed Oct, 10, 2015).

National Weather Service Weather Forecast Office. "The Black Sunday Dust Storm of 14 April 1935." http://www.srh.noaa.gov/oun/?n=events-19350414 (accessed Oct. 23, 2015).

The Nature Conservancy. "Kids These Days: Why Is America's Youth Staying Indoors." http://www.nature.org/newsfeatures/kids-in-nature/kids-in-nature-poll.xml (accessed Oct. 9, 2015).

North American Bird Conservation Initiative, U.S. Committee. *The State of the Birds 2014 Report*. Washington, DC: U.S. Department of Interior, 2014. 16 pages. http://www.stateofthebirds.org (accessed Aug. 18, 2015).

Nutter, Felicia Beth. "Evaluation of a trap-neuter-return management program for feral cat colonies: Population dynamics, home ranges, and potentially zoonotic diseases." PhD diss., North Carolina State University, Raleigh, 2006.

O'Brien, Michael, Richard Crossley, and Kevin Karlson. *The Shorebird Guide*. New York: Houghton Mifflin Company, 2006.

Pacelle, Wayne. "Finding Common Ground: Outdoor Cats and Wildlife." *A Humane Nation: Wayne Pacelle's Blog*, Nov. 21, 2011. http://blog.humanesociety.org/wayne/2011/11/feral-cats-wildlife.html?credit=blog_post (accessed Sep. 20, 2015).

Palanisamy, Manikandan, Bhaskar Madhavan, Manohar Babu Balasundaram, Raghuram Andavar, and Narendran Venkatapathy. "Outbreak of ocular toxoplasmosis in Coimbatore, India." *Indian Journal of Ophthalmology* 54, no. 2 (2006): 129.

Patronek, G. J. "Free-roaming and feral cats—their impact on wildlife and human beings." *Journal of the American Veterinary Medical Association* 212, no. 2 (1998): 218–26.

People for the Ethical Treatment of Animals. "The Deadly Consequences of 'No-Kill' Policies?" http://www.peta.org/features/deadly-consequences-no-kill-policies/ (accessed Oct. 14, 2015).

People for the Ethical Treatment of Animals. "PETA's History: Compassion in Action." http://www.peta.org/about-peta/learn-about-peta/history/.

People for the Ethical Treatment of Animals. "What is PETA's stance on programs that advocate trapping, spaying and neutering, and releasing feral cats?" http://www.peta.org/about-peta/faq/what-is-petas-stance-on-programs-that-advocate-trapping-spaying-and-neutering-and-releasing-feral-cats.

Peterson, Roger Tory. *Peterson Field Guide to Birds of North America*. New York: Houghton Mifflin Harcourt, 2008.

Pet Food Institute. "Pet Food Sales." http://www.petfoodinstitute.org/?page=PetFoodSales (accessed Jul. 6, 2015).

Ramzy, Austin. "Australia Writes Morrissey to Defend Plan to Kill Millions of Feral Cats." *New York Times*, Oct. 14, 2015. http://www.nytimes.com/2015/10/15/world/australia/australia-feral-cat-cull-brigitte-bardot-morrissey.html (accessed Oct. 20, 2015).

Ratcliffe, Norman, Mike Bell, Tara Pelembe, Dave Boyle, Raymond Benjamin Richard White, Brendan Godley, Jim Stevenson, and Sarah Sanders. "The eradication of feral cats from Ascension Island and its subsequent recolonization by seabirds." *Oryx: The International*

Journal of Conservation (published for Fauna and Flora International) no. 44, 1 (2009): 20–29.

Raup, D., and J. Sepkoski Jr. "Mass extinctions in the marine fossil record." *Science* 215, 4539 (1982): 1501–3.

Recuenco, Sergio, Bryan Cherry, and Millicent Eidson. "Potential cost savings with terrestrial rabies control." *BMC Public Health* 7, no. 1 (2007): 47.

Renne, Paul R., Alan L. Deino, Frederik J. Hilgen, Klaudia F. Kuiper, Darren F. Mark, William S. Mitchell, Leah E. Morgan, Roland Mundil, and Jan Smit. "Time Scales of Critical Events Around the Cretaceous-Paleogene Boundary." *Science* 339, no. 6120 (Feb. 7, 2013): 684–87.

Rice, Harvey. "Galveston Bird Watcher/Cat Killer Won't Be Retried." *Houston Chronicle*, Nov. 16, 2007. http://www.chron.com/news/houston-texas/article/Galveston-bird-watcher-cat-killer-won-t-be-retried-1647458.php.

Rocus, Denise S., and Frank Mazzotti. "Threats to Florida's Biodiversity." University of Florida IFAS Extension. http://edis.ifas.ufl.edu/uw107 (accessed Oct. 7, 2015).

Roebling, A. D., D. Johnson, J. D. Blanton, M. Levin, D. Slate, G. Fenwick, and C. E. Rupprecht. "Rabies Prevention and Management of Cats in the Context of Trap-Neuter-Vaccinate-Release Programmes." *Zoonoses and Public Health* 61, 4 (2014): 290–96.

Roelke, M. E., D. J. Forester, E. R. Jacobson, G. V. Kollias, F. W. Scott, M. C. Barr, J. F. Evermann, and E. C. Pirtel. "Seroprevalence of infectious disease agents in free-ranging Florida panthers (*Felis concolor coryi*)." *Journal of Wildlife Diseases* 29 (1993): 36–49.

Royal Society for the Protection of Birds. "Are Cats Causing Bird Declines?" http://www.rspb.org.uk/makeahomeforwildlife/advice/gardening/unwantedvisitors/cats/birddeclines.aspx. (accessed Apr. 18, 2015).

Royal Society for the Protection of Birds. "Decline of Urban House Sparrows." http://www.rspb.org.uk/whatwedo/projects/details/198323-causes-of-population-decline-of-urban-house-sparrows- (accessed Apr. 18, 2015).

Rupprecht, Charles E., Cathleen A. Hanlon, and Thiravat Hemachudha. "Rabies re-examined." *The Lancet Infectious Diseases* 2, no. 6 (2002): 327–43.

Santella, Chris. *Fifty Places to Go Birding Before You Die*. New York: Stewart, Tabori & Chang, 2007.

Schopf, J. W., A. B. Kudryavtsev, A. D. Czaja, and A. B. Tripathi. "Evidence of Archean Life: Stromatolites and Microfossils." *Precambrian Research* 158 (2007): 141–155.

Seton, Ernest Thompson. *Two Little Savages: Being the Adventures of Two Boys Who Lived as Indians*. Oxford: Benediction Classics, 2008.

Sleeman, J. M., J. M. Keane, J. S. Johnson, R. J. Brown, and S. V. Woude. "Feline leukemia virus in a captive bobcat." *Journal of Wildlife Diseases* 37, 1 (2001): 194–200.

Stallcup, R. "Cats: A Heavy Toll on Songbirds. A Reversible Catastrophe." Focus 29. *Quarterly Journal of the Point Reyes Bird Observatory* (Spring/Summer 1991): 8–9. http://www.pointblue.org/uploads/assets/observer/focus29cats1991.pdf.

Stanley Medical Research Institute. "Toxoplasmosis–Schizophrenia Research." http://www.stanleyresearch.org/patient-and-provider-resources/toxoplasmosis-schizophrenia-research/ (accessed Sep. 19, 2015).

State of New York, Department of Agriculture. Article 26 of the Agriculture and Markets Law Relating to Cruelty to Animals. http://www.agriculture.ny.gov/ai/AILaws/Article_26_Circ_916_Cruelty_to_Animals.pdf.

Stearns, Beverly Peterson, and Stephen C. Stearns. *Watching, from the Edge of Extinction.* New Haven, CT: Yale University Press, 2000.

Stenseth, Nils Chr., Bakyt B. Atshabar, Mike Begon, Steven R. Belmain, Eric Bertherat, Elisabeth Carniel, Kenneth L. Gage, Herwig Leirs, and Lila Rahalison. "Plague: Past, present, and future." *PLoS Med* 5, no. 1 (2008): e3.

Stenseth, Nils Chr., Noelle I. Samia, Hildegunn Viljugrein, Kyrre Linné Kausrud, Mike Begon, Stephen Davis, Herwig Leirs, et al. "Plague dynamics are driven by climate variation." *Proceedings of the National Academy of Sciences* 103, no. 35 (2006): 13110–15.

Tharoor, Ishaan. "Australia actually declares 'war' on cats, plans to kill 2 million by 2020." *Washington Post*, Jul. 16, 2015. https://www.washingtonpost.com/news/worldviews/wp/2015/07/16/australia-actually-declares-war-on-cats-plans-to-kill-2-million-by-2020/.

Theodore Roosevelt Association. "The Conservationist." http://www.theodoreroosevelt.org/site/pp.aspx?c=elKSIdOWIiJ8H&b=8344385 (accessed Aug. 16, 2015).

Tobin, Kate. *The Rundown.* "Did wolves help restore trees to Yellowstone?" Sep. 4, 2015. http://www.pbs.org/newshour/rundown/wolves-greenthumbs-yellowstone.

Torrey, E. F., J. J. Bartko, and R. H. Yolken. "*Toxoplasma gondii*: Meta-analysis and assessment as a risk factor for schizophrenia." *Schizophrenia Bulletin* 38 (2012): 642–47.

Torrey, E. Fuller, and Robert H. Yolken. "*Toxoplasma gondii* and schizophrenia." *Emerging Infectious Diseases* 9, no. 11 (2003): 1375.

United States Census Bureau. "America's Families and Living Arrangements: 2012." Aug. 2013. http://www.census.gov/prod/2013pubs/p20-570.pdf.

United States Department of Agriculture. Natural Resources Conservation Service. "Hugh Hammond Bennett: 'Father of Soil Conservation.'" http://www.nrcs.usda.gov/wps/portal/nrcs/detail/national/about/history/?cid=stelprdb1044395 (accessed Oct. 23, 2015).

U.S. Fish and Wildlife Service. "Birding in the United States: A Demographic and Economic Analysis." http://www.fws.gov/southeast/economicImpact/pdf/2011-BirdingReport--FINAL.pdf (accessed Apr. 20, 2015).

U.S. Fish and Wildlife Service. "Digest of Federal Resource Laws of Interest to the U.S. Fish and Wildlife Service." http://www.fws.gov/laws/lawsdigest/esact.html (accessed Jul. 22, 2015).

U.S. Fish and Wildlife Service. "Endangered Species Act | Section 3." http://www.fws.gov/endangered/laws-policies/section-3.html.

U.S. Fish and Wildlife Service. "Piping Plover Fact Sheet." http://www.fws.gov/midwest/endangered/pipingplover/pipingpl.html (accessed Jul. 20, 2015).

Velasco-Murgía, M. *Colima y las islas de Revillagigedo*. Colima, Mexico: Universidad de Colima, 1982.

Vuilleumier, François. "Dean of American Ornithologists: The Multiple Legacies of Frank M. Chapman of the American Museum of Natural History." *The Auk* 122, no. 2 (2005): 389–402.

Vyas, Ajai, Seon-Kyeong Kim, Nicholas Giacomini, John C. Boothroyd, and Robert M. Sapolsky. "Behavioral changes induced by *Toxoplasma* infection of rodents are highly specific to aversion of cat odors." *Proceedings of the National Academy of Sciences* 104, no. 15 (2007): 6442–47.

Warner, R. E. "Demography and movements of free-ranging domestic cats in rural Illinois." *Journal of Wildlife Management* 49, 2 (1985): 340–46.

Watts, E., Y. Zhao, A. Dhara, B. Eller, A. Patwardhan, and A. P. Sinai. "Novel approaches reveal that *Toxoplasma gondii* bradyzoites within tissue cysts are dynamic and replicating entities in vivo." *MBio* 6, no. 5 (2015): e01155-15.

Williams, Ted. "The Most Endangered Bird in the Continental U.S." *Audubon*, Mar.–Apr. 2013. https://www.audubon.org/magazine/march-april-2013/the-most-endangered-bird-continental-us (accessed Oct. 24, 2015).

Wilson, Don E., and DeeAnn M. Reeder (eds). *Mammal Species of the World: A Taxonomic and Geographic Reference*. 3rd ed. Baltimore: Johns Hopkins University Press, 2005.

Winter, L. "Popoki and Hawaii's Native Birds." *'Elepaio* 63 (2003): 43–46.

Winter, L. "Trap-neuter-release programs: The reality and impacts." *Journal of the American Veterinary Medical Association* 225 (2004): 1369–76.

Wisch, Rebecca F. "Detailed Discussion of State Cat Laws." Michigan State University College of Law. https://www.animallaw.info/article /detailed-discussion-state-cat-laws (accessed Jul. 15, 2015).

Work, Thierry M., J. Gregory Massey, Bruce A. Rideout, Chris H. Gardiner, David B. Ledig, O. C. H. Kwok, and J. P. Dubey. "Fatal toxoplasmosis in free-ranging endangered 'Alalā from Hawaii." *Journal of Wildlife Diseases* 36, no. 2 (2000): 205–12.

World Health Organization. "Rabies Fact Sheet No. 99." Mar. 2016. http://www.who.int/mediacentre/factsheets/fs099/en/ (accessed Apr. 7, 2016).

World History of Art. "Cats in History." http://www.all-art.org/Cats/BIG _BOOK1.htm (accessed Apr. 4, 2015).

Zasloff, Lee R., and Lynette A. Hart. "Attitudes and care practices of cat caretakers in Hawaii." *Anthrozoös* 11, no. 4 (1998): 242–48.

INDEX

Page numbers in **bold** refer to illustrations found in the plates section